# Careers in Cybersecurity

# Careers in Cybersecurity

SALEM PRESS
A Division of EBSCO Information Services, Inc.
Ipswich, Massachusetts

**GREY HOUSE PUBLISHING**

Cover photo: iStock



*Careers in Cybersecurity*, published by Grey House Publishing, Inc., Amenia, NY, under exclusive license from EBSCO Information Services, Inc.



∞ The paper used in these volumes conforms to the American National Standard for Permanence of Paper for Printed Library Materials, Z39.48 1992 (R2009).

Publisher's Cataloging-in-Publication Data
(Prepared by Parlew Associates, LLC)

Title: Careers in cybersecurity.
Description: Ipswich, MA : Salem Press, a division of EBSCO Information Services, Inc. ; Amenia, NY : Grey House Publishing, 2023. | Series: [Careers in ]. | Includes bibliographical references and index. | Includes b&w photos.
Identifiers: ISBN 9781637004128 (hardback) | ISBN 9781637004135 (ebook)
Subjects: LCSH: Computer networks -- Security measures — Vocational guidance -- United States. | Computer security -- Vocational guidance — United States. | BISAC: BUSINESS & ECONOMICS / Industries / Computers & Information Technology. | COMPUTERS / Security / General.
Classification: LCC TK5105.59 2023 (print) | LCC TK5105.59 (ebook) | DDC 005.8023—dc23

First Printing

PRINTED IN THE UNITED STATES OF AMERICA

# CONTENTS

# PUBLISHER'S NOTE

The latest volume in our *Careers in* series, *Careers in Cybersecurity*, provides a thorough introduction to the wide range of career opportunities available to those interested in the many areas of this vital industry, from computer programming to financial management, and everything in between.

The types of careers related to the cybersecurity industry are incredibly wide-ranging, from the obvious, like information security analyst, to the not-so-obvious, like loan officer or tax examiner. *Careers in Cybersecurity* offers insight into what it's like to work behind the scenes at a cybersecurity firm, investigate cybercrime from a law-enforcement perspective, and utilize cybersecurity measures to protect sensitive data. If you are interested in how to turn your passion for computers and technology into a fulfilling career, this book is for you.

At the heart of this volume, like the dozens of previous *Careers in* titles, are 28 occupation profiles that cover various careers involving cybersecurity in some way, from Computer Information Research Scientist to Forensic Science Technician to Software Developer. Each career profile offers details about its specific career:

**Snapshots** offer current data on:

- Median pay
- Job outlook
- Educational and other requirements
- Working conditions
- Recommended areas of interest

**Career Overviews** describe:

- Duties
- Work environment
- Work schedules

In addition, each profile provides important qualities for the job, specific and general job requirements, anticipated rates of growth, and why. Also included is a list of related occupations, both in the public and private sectors, and a list of resources for more information, with profession-specific associations and certifying bodies.

Merging scholarship with occupational development, this single comprehensive guidebook provides students who are interested in exploring a career involving cybersecurity with the necessary insight into the wide array of options available in this highly skilled and ever-evolving field. The book offers guidance regarding what job-seekers can expect in terms of training, advancement, earnings, job prospects, working conditions, relevant associations, and more. *Careers in Cybersecurity* is specifically designed for a high school and undergraduate audience and is edited to align with secondary or high school curriculum standards.

**Additional Materials**

*Careers in Cybersecurity* is enhanced with statistical data, including projections from the U.S. Bureau of Labor Statistics, and median annual salaries or wages for those occupations profiled. Fast Facts, Famous Firsts, and dozens of photos add depth to the discussion.

A popular and valuable highlight of the *Careers in* series are "Conversations With..." —interviews with real-life professionals working in relevant jobs. Those interviewed offer their personal career paths, detail the potential for career advancement, and offer advice for students.

The back matter in this volume includes:

Appendix A: Guide to the Holland Code—John Holland's theory that people and work environments can be classified into six different groups: realistic, investigative, artistic, social, enterprising, and conventional. These interest codes are found at the beginning of each profile.

Appendix B: General Bibliography—a list of suggested readings for those who wish to learn more about pursuing a career in artificial intelligence.

Appendix C: List of Organizations and Resources—a list of additional resources for education, certification, professional development, and networking.

Salem Press thanks the professionals whose honest responses and personal insight provide immeasurable value to *Careers in Cybersecurity.*

# INTRODUCTION

At no other time in history has humanity been so interconnected. Computers and the internet power our daily lives, from the moment we wake up to the chime of a smartphone, to the actions we perform at work, to the moment we fall asleep at night while binge-watching the latest streaming series. With the prevalence of and rapid advances in digital technology, there is perhaps no industry of greater importance today than cybersecurity.

Experts in the cybersecurity field work tirelessly behind the scenes to keep our way of life running smoothly, in all sectors: healthcare, education, finance, media, government, military, and more. The list of careers that are related to cybersecurity is long, including everything from Accountants and Auditors to Tax Examiners, Collectors and Revenue Agents. Our goal with this edition is not to be all-encompassing, but rather to pinpoint key careers in this important and constantly evolving industry.

Digital technology has come a long way since Charles Babbage invented the first automatic digital computer in the 1820s—a mechanical calculator called a "difference engine," that could tabulate polynomial functions. Today, the world's fastest super-computer (the Frontier) carries out functions at 1.102 quintillion operations per second. It has been said that technology advances exponentially—for instance, in less than 20 years, society has gone from reconnecting with old friends on Facebook to anticipating the next leaps in the immersive metaverse.

With this incredible advancement in technology comes increasing integration with all facets of human life, meaning security is of paramount importance. Cybercriminals may exploit vulnerabilities for profit, or merely to cause chaos. As technology and inherent threats progress, so do the opportunities for skilled individuals to enter the arena of cybersecurity.

Finding the path that's right for you means thinking about your own interests, skills, and talents. Important skills for working in cybersecurity include analysis, communication, logic and problem-solving, and math skills, not to mention a love of computers and technology in general.

It's also worth considering how much money you want to make. A position that requires a four-year degree will earn you more money than one requiring a two-year degree, or one requiring no post-secondary education at all. The profiles in this volume include education requirements and typical earnings. Remember, however, that how much money you make is not the entire story. Benefits, job security, where you work, and self-fulfillment are important factors as well.

The "Conversations With..." spread throughout this volume show the variety of career paths available the cybersecurity field, how to achieve them, and what to expect when you get there. They are interviews with real individuals working in the field at real jobs.

The list of jobs contained in this volume is not exhaustive, and should be viewed as an entry point into a world with numerous branches and sub-disciplines, many of which can intersect in novel ways. The aim is to provide readers—especially students embarking on their lifelong careers—with accurate and detailed examples of some the many possibilities available in this industry, which is so vital to the functioning of our modern world.

Here are details about how certain careers can relate to cybersecurity:

*Accountants and auditors* prepare and examine financial records, identify potential areas of opportunity and risk, and provide solutions for businesses and individuals. In accounting, cybersecurity is of ever-increasing importance, as more firms manage and store valuable financial data and sensitive information digitally, thus heightening the risk of attacks by cybercriminals.

*Computer and information research scientists* design innovative uses for new and existing technology. They study and solve complex problems in computing for business, science, medicine, and other fields. A key problem these research scientists face is the need to develop new and innovative cybersecurity measures to fend off attacks from cybercriminals whose methods are always evolving.

*Computer programmers* write, modify, and test code and scripts that allow computer software and applications to function properly. They turn the designs created by software developers and engineers into instructions that a computer can follow. In addition, programmers run tests to ensure that newly created applications and software produce the expected results. If the products do not work correctly, programmers check the code or scripts for mistakes and modify them. This work can be exceedingly important, depending on the nature of the program being written, as in the case of cybersecurity programs.

*Forensic science technicians* aid criminal investigations by collecting and analyzing evidence. Digital forensics involves the gathering, analyzation and securing of digital evidence for forensic investigative purposes.

*Information security analysts* plan and carry out security measures to protect an organization's computer networks and systems. As the number of organizations that do business online increases, so do the chances of cyberthreats increase. It is up to these analysts to create innovative solutions to prevent hackers from stealing critical information or creating problems for computer networks.

*Mathematicians and statisticians* analyze data and apply computational techniques to solve problems. They typically work in any field that benefits from data analysis, including education, government, healthcare, and research and development. In the latter field, they may work on teams attempting to advance technology such as the latest cybersecurity measures.

*Police officers* protect lives and property. Detectives and criminal investigators, who are sometimes called agents or special agents, gather facts and collect evidence of crimes.

With the increasing hold the internet and wireless networks have over our daily lives, it is no surprise that many police departments have units dedicated to the investigation of cyber related crimes, where some actions undertaken can include the identification of places (GPS tracking of suspects; pinpointing potential crime scenes) and persons (witnesses, suspects), email and IP tracing, etc.

*Tax examiners, collectors, and revenue agents* determine how much is owed in taxes and collect tax from individuals and businesses on behalf of federal, state, and local governments. Because of the sensitive financial information they deal in, it is a top concern for these workers to protect their systems from crimes such as ransomware, spyware, computer viruses, phishing, vishing, spoofing, data breaches, and more.

# Accountant/Auditor

## Snapshot

**Career Cluster(s):** Business Management & Administration; Finance; Law, Public Safety, Corrections & Security

**Interests:** Accounting; Finance; Data; Statistics; Recordkeeping; Investigating

**Earnings (Yearly Average):** $77,250

**Employment & Outlook:** As Fast As Average Growth Expected

## OVERVIEW

### Sphere of Work

Accountants and auditors prepare and examine financial records, identify potential areas of opportunity and risk, and provide solutions for businesses and individuals. They ensure that financial records are accurate, that financial and data risks are evaluated, and that taxes are paid properly. They also assess financial operations and work to help ensure that organizations run efficiently. In accounting, cybersecurity is of ever-increasing importance, as more firms manage and store valuable financial data and sensitive information digitally, thus heightening the risk of attacks by cybercriminals.

### Work Environment

Accountants work in offices or at home. Some work on teams, whereas others work alone. Most accountants and auditors work full time. Longer periods of work are typical at certain times of the year, such as for quarterly audits or during tax season.

Photo via iStock.com/KangeStudio. [Used under license.]

## Occupation Interest

Accountants and auditors have strong mathematical skills and an interest in finances and meticulous work like recordkeeping. They are highly organized and patient. Accountants must also have an added interest in structuring a cybersecurity strategy to keep financial data secure.

### Profile

**Working Conditions:** Inside
**Education Needs:** Bachelor's Degree; Master's Degree
**Licensure/Certification:** Various
**Opportunities for Experience:** Internship
**Interest Score:** CEI

## A Day in the Life—Duties and Responsibilities

### Duties and Responsibilities

- Examining financial statements to ensure that they are accurate and comply with laws and regulations
- Computing taxes owed, preparing tax returns, and ensuring that taxes are paid properly and on time
- Inspecting account books and accounting systems for efficiency and use of accepted accounting procedures and identifying potential risks for fraud
- Organizing, analyzing, and maintaining financial records
- Assessing financial operations, identifying risks and challenges, and making best-practices recommendations to management
- Suggesting ways to reduce costs, enhance revenues, and improve profits

In addition to examining and preparing financial documents, accountants and auditors must explain their findings. This includes preparing written reports and meeting face-to-face with organization managers and individual clients.

Many accountants and auditors specialize, depending upon their employer. Some work for organizations that specialize in assurance services (improving the quality or context of information for decision makers) or risk management (determining the probability of a misstatement on financial documents). Other organizations specialize in specific industries, such as finance, insurance, or healthcare.

## OCCUPATION SPECIALTIES

### External Auditor

External auditors check for proper management of an organization's funds, sources of revenue, and internal controls, such as financial data preparation or managing risks to cybersecurity or the supply chain. They are employed by an outside organization, rather than the one they are auditing. They review clients' financial statements and inform authorities, investors, and regulators that the statements have been correctly prepared and reported with no material misstatements.

**Government Accountant**

Government accountants maintain and examine the records of government agencies and audit private businesses and individuals whose activities are subject to government regulations or taxation. Accountants employed by federal, state, and local governments ensure that revenues are received and spent according to laws and regulations. Their responsibilities include auditing, financial reporting, and management accounting.

**Information Technology (IT) Auditor**

Information technology (IT) auditors review controls for their organization's IT systems to ensure that both financial and nonfinancial data come from a reliable source.

**Internal Auditor**

Internal auditors have duties that are similar to external auditors, but these workers are employed by the organization they are auditing. They identify ways to improve the processes for finding and eliminating waste, fraud, and other financial risks to the organization. The practice of internal auditing is not regulated, but the Institute of Internal Auditors (IIA) provides generally accepted standards.

**Management Accountant**

Management accountants are also called cost, corporate, industrial, managerial, or private accountants. They combine accounting and financial information to guide business decision making. They also understand financial and nonfinancial data and how to integrate information. The information that management accountants prepare is intended for internal use by business managers, not for the public.

Management accountants often prepare budgets and evaluate performance. They also may help organizations plan the cost of doing business. Some work with financial managers on asset management, which involves planning and selecting financial investments such as stocks, bonds, and real estate.

**Public Accountant**

Public accountants have a broad range of accounting, auditing, tax, and consulting tasks. Their clients include corporations, governments, individuals, and nonprofits.

Public accountants work with financial documents that clients are required by law to disclose, such as tax forms and financial statements that corporations must provide to current and potential investors. Some public accountants concentrate on tax matters, advising corporations about the tax advantages of certain business decisions or preparing individual income tax returns.

Forensic accountants fall under this category. They often work closely with law enforcement personnel and lawyers during investigations and may also appear as expert witnesses during trials.

Still others collaborate with individuals, advising them on important personal financial matters. These public accountants combine their expertise in data management, economics, financial planning, and tax law to develop strategies for their clients. Advisory services cover topics including cash flow, insurance, investment, retirement, and wealth transfer planning to help clients meet financial goals, such as retirement, paying for a child's education, or buying a home.

Public accountants, many of whom are Certified Public Accountants (CPAs), generally have their own businesses or work for public accounting firms. Publicly traded companies are required to have CPAs sign documents they submit to the Securities and Exchange Commission (SEC), including annual and quarterly reports.

## WORK ENVIRONMENT

### Immediate Physical Environment

Most accountants and auditors work in offices, but some work from home. Although accountants and auditors usually work in teams, some work alone. Accountants and auditors may travel to their clients' places of business.

### Human Environment

Accountants work for clients by the very nature of their job, so must be comfortable interacting with people in person, on the phone, and by email. Accountants should be professional and accommodating to their clients' needs. They may also be required to present reports to management.

### Technological Environment

Accountants and auditors may use technology, such as artificial intelligence (AI) and robotics process automation (RPA) to increase their productivity. Automating some routine tasks makes these workers more efficient by allowing them to focus on analysis and other high-level responsibilities.

## EDUCATION AND TRAINING

### High School/Secondary

High school students interested in becoming accountants or auditors should focus on math, statistics, and business-related courses. Those interested in forensic accounting specifically should also include courses such as civics and psychology. Becoming an accountant or auditor requires at least a bachelor's degree, so students should prepare themselves for postsecondary education, including researching the schools they would like to consider attending to achieve that goal.

**Suggested High School Subjects**

- Algebra
- Biology
- Chemistry
- Civics
- Earth or Life or Physical Science
- Economics
- English
- Geography
- Geometry
- History
- Physics
- Pre-Calculus
- Psychology
- Statistics
- Trigonometry

**Related Career Pathways/Majors**

*Business Management & Administration Career Cluster*

- General Management Pathway
- Operations Management Pathway

*Finance Career Cluster*

- Accounting Pathway
- Business Finance Pathway

*Law, Public Safety, Corrections & Security Career Cluster*

- Law Enforcement Services Pathway
- Legal Services Pathway

***Transferable Skills and Abilities***

**Analytical and Critical-thinking Skills**

- Critically evaluating data, identifying issues in documentation, and suggesting solutions.

**Communication Skills**

- Listening to and discussing facts and concerns from clients, managers, and other stakeholders
- Discussing the results of their work both in meetings and in written reports

**Detail-oriented**

- Paying attention to details when compiling and examining documents

**Math Skills**

- Analyzing, comparing, and interpreting facts and figures
- Using advanced math skills, such as calculus and statistical analysis

**Organizational Skills**

- Working with a range of financial documents for a variety of clients

## Postsecondary

Most accountant and auditor positions require at least a bachelor's degree in accounting or a related field. Some employers prefer to hire applicants who have a master's degree, either in accounting or in business administration with a concentration in accounting.

Some universities and colleges offer specialized programs for a bachelor's or master's degree, such as in accounting, forensic accounting, internal auditing, or tax accounting. In some cases, those with an associate degree, as well as bookkeepers, accounting, and auditing clerks who meet the education and experience requirements set by their employers, may get junior accounting positions and advance by exhibiting their accounting skills on the job.

Students may gain practical experience through internships with public accounting or business firms.

### Related College Majors

- Accounting
- Auditing
- Business Management & Administration
- Economics
- Entrepreneurial Studies
- Financial Management
- General Business
- Insurance
- International Business
- Marketing

## Adult Job Seekers

Adults seeking to become professional accountants or auditors should have a background or strength in math or finance. Schooling is still required, so thought should be given to an individual's long-term professional goals. An adult seeking to become a forensic accountant may also have a background in law enforcement.

## Professional Certification and Licensure

CPAs are licensed by their state's Board of Accountancy. Becoming a CPA requires passing a national exam and meeting other state requirements. All states require CPA candidates to complete 150 semester hours of college coursework to be licensed, which is 30 hours more than the usual 4-year bachelor's degree.

Many schools offer a 5-year combined bachelor's and master's degree to meet the 150-hour requirement, but a master's degree is not required.

A few states allow a number of years of public accounting experience to substitute for a college degree. For the experience to count, it must be performed under the direct supervision and oversight of a CPA with an active license.

All states use the four-part Uniform CPA Examination from the American Institute of Certified Public Accountants (AICPA). Candidates do not have to pass all four parts at once, but most states require that candidates pass all four parts within 18 months of passing their first part.

All states require CPAs to take continuing education courses, including ethics, to maintain their license.

Certification provides an advantage in the job market because it shows professional competence in a specialized field of accounting and auditing. Accountants and auditors seek certifications from a variety of professional societies. Some of the most common certifications are listed below:

- The AICPA offers several designations. For accountants with a CPA, the AICPA offers the Accredited in Business Valuation (ABV), Certified Financial Forensics (CFF), Certified Information Technology Professional (CITP), and Personal Financial Specialist (PFS) certifications. All these credentials require experience in the related area, continuing education, and passing an exam.

- AICPA and the Chartered Institute of Management Accountants (CIMA) developed the Chartered Global Management Accountant (CGMA) designation as an internationally recognized professional credential. Candidates must complete a program, pass an exam, and meet a requirement for work experience.

- The Association of Government Accountants (AGA) offers the Certified Government Financial Manager (CGFM) credential to accountants or auditors working with federal, state, or local government. To earn this certification, candidates must have a bachelor's degree from an accredited college or university, pass examinations, and have professional-level experience in government financial management. To keep the certification, CGFMs must complete continuing professional education.

- The Institute of Internal Auditors (IIA) offers the Certified Internal Auditor (CIA) credential to graduates from accredited colleges and universities who have work experience as internal auditors and have passed an exam. The IIA also offers the Certified in Control Self-Assessment (CCSA), Certified Government Auditing Professional (CGAP), Certified Financial Services Auditor (CFSA), and Certification in Risk Management Assurance (CRMA) to those who pass the exams and meet educational and experience requirements.

- The Institute of Management Accountants (IMA) offers the Certified Management Accountant (CMA) to applicants who complete a bachelor's degree. Applicants must have work experience in management accounting, pass an exam, agree to meet continuing education requirements, and comply with standards of professional conduct.

- Information Systems Audit and Control Association (ISACA) offers the Certified Information Systems Auditor (CISA) to candidates who pass an exam and have work experience auditing information systems. Information systems experience, financial or operational auditing experience, or related college credit hours may be substituted for some of the experience required in information systems auditing, control, or security.

## EARNINGS AND ADVANCEMENT

Earnings depend upon the area of accounting or auditing being practiced, and whether the individual is self-employed or works for a firm. Median annual earnings of accountants and auditors were $77,250 in 2021. The lowest 10 percent earned less than $47,970, and the highest 10 percent earned more than $128,970.

Accountants and auditors may be able to expense travel to locations during audits or investigations, for example.

Entry-level public accountants may advance to senior positions as they gain experience and take on more responsibility. Those who excel may become supervisors, managers, or partners; open their own public accounting firm; or transfer to executive positions in management accounting or internal auditing in private firms.

Management accountants often start as cost accountants, junior internal auditors, or trainees for other accounting positions. As they rise through the organization, they may advance to accounting managers, budget directors, chief cost accountants, or managers of internal auditing. Some become controllers, treasurers, financial vice presidents, chief financial officers, or corporation presidents.

Public accountants, management accountants, and internal auditors may move from one type of accounting and auditing to another. Public accountants often move into management accounting or internal auditing. Management accountants may become internal auditors, and internal auditors may become management accountants. However, it is less common for management accountants or internal auditors to move into public accounting.

## EMPLOYMENT AND OUTLOOK

Accountants and auditors held 1,449,800 jobs in 2021. Employment is expected to grow as fast as average for all occupations through the year 2031, at a rate of 4 percent. About 136,400 openings for accountants and auditors are projected each year, on average, over the decade.

Globalization, a growing economy, and a complex tax and regulatory environment are expected to continue to lead to strong demand for accountants and auditors.

In general, employment growth of accountants and auditors is expected to be closely tied to the health of the overall economy. As the economy grows, these workers will continue to be needed to prepare and examine financial records. In addition, as more companies go public, there will be greater need for public accountants to oversee the legally required financial documentation.

The continued globalization of business may lead to increased demand for accounting expertise and services related to international trade and international mergers and acquisitions.

Technological change is expected to affect the role of accountants over the decade. Some routine accounting tasks may be automated as platforms such as cloud computing, artificial intelligence (AI), and blockchain become more widespread. Although this will allow accountants to become more efficient, this change is not expected to reduce overall demand. Instead, with the automation of routine tasks, such as data entry, the advisory and analytical duties of accountants will become more prominent.

## Famous First

**The vacuum tube, which served as the foundation for IBM's first major computer when it was unveiled in 1952, was affordable enough for businesses to purchase, making accountants among the first professionals to use it. Pictured is the IBM 701 console, c. 1950s.**

Source: investopedia.com

Photo by Dan, via Wikimedia Commons.

Many openings are expected to result from the need to replace workers who transfer to different occupations or exit the labor force, such as to retire. Demand for accountants may lead to good prospects for entry-level positions. However, competition will be strong for jobs with the most prestigious accounting and business firms.

Accountants and auditors who have earned professional recognition, especially as CPAs, should have the best prospects. Job applicants who have a master's degree in accounting or a Master of Business Administration (MBA) degree with a concentration in accounting also may have an advantage.

### Related Occupations

- Bookkeeping/Accounting/Auditing Clerks
- Budget Analyst
- Cost Estimator
- Financial Analyst
- Financial Manager
- Management Analyst
- Personal Financial Advisor
- Postsecondary Teacher
- Tax Examiner/Collector/Revenue Agent
- Top Executive

## MORE INFORMATION

**American Accounting Association (AAA)**
9009 Town Center Parkway
Lakewood Ranch, FL 34202
941.921.7747
info@aaahq.org
aaahq.org
Forensic Accounting Division: aaahq.org/FIA

**American Board of Forensic Accounting (ABFA)**
1400 Easton Drive, Suite 137
Bakersfield, CA 93309
661.861.8861
staff@abfa.us

**Association of Certified Fraud Examiners (ACFA)**
Global Headquarters—The Gregor Building
716 West Avenue
Austin, TX 78701-2727
800.245.3321
MemberServices@ACFE.com
www.acfe.com

**Association of Government Accountants (AGA)**
2208 Mount Vernon Avenue
Alexandria, VA 22301-1314
703.684.6931
www.agacgfm.org

**Association of International Certified Professional Accountants (AICPA)**
220 Leigh Farm Rd.
Durham, NC 27707-8110
888.777.7077
www.aicpa.org

**Association to Advance Collegiate Schools of Business (AACSB)**
777 S. Harbour Island Boulevard, Suite 750
Tampa, FL 33602
813.769.6500
www.aacsb.edu

**Forensic Accountants Society (FASNA)**
6200 Aurora Avenue, Suite 402W
Urbandale, IA 50322-2800
515.276.7000
www.fasna.org

**Forensic Certified Public Accountants Association (FCPAA)**
1120 Avenue of the Americas, Room 4057, 4th Floor
New York, NY 10036
626.521.9077
info@fcpachina.org
www.fcpai.org

**Forensic CPA Society**
P.O. Box 31060
Spokane, WA 99223
509.448.9318
www.fcpas.org

**Global Academy of Finance and Management (GAFM)**
P.O. Box 293
1670F E. Cheyenne Mountain Boulevard
Colorado Springs, CO 80906
www.gafm.com

**Information Systems Audit and Control Association (ISACA)**
1700 E. Golf Road, Suite 400
Schaumburg, IL 60173
847.253.1545
www.isaca.org

**Institute of Certified Forensic Accountants (ICFA)**
3422 Old Capitol Trail, Suite 1063
Wilmington, DE 19808-6192
302.425.5528
www.forensicglobal.org

**Institute of Internal Auditors (IIA)**
1035 Greenwood Blvd., Suite 401
Lake Mary, FL 32746
407.937.1111
na.theiia.org

**Institute of Management Accountants (IMA)**
10 Paragon Drive, Suite 1
Montvale, NJ 07645-1760
800.638.4427
ima@imanet.org
www.imanet.org

**National Association of Forensic Accountants (NAFA)**
6635 West Commercial Boulevard, Suite 217
Tamarac, FL 33319
844.217.6975
mail@nafanet.com
www.nafanet.com

---

*Stuart Paterson; updated by Daniel Vest*

# Budget Analyst

## Snapshot

**Career Cluster(s):** Business Management & Administration; Finance

**Interests:** Accounting; Finance; Data; Statistics; Recordkeeping; Investigating

**Earnings (Yearly Average):** $79,940

**Employment & Outlook:** Slower Than Average Growth Expected

## OVERVIEW

### Sphere of Work

Budget analysts help public and private organizations plan their finances. They prepare budget reports and monitor organizational spending. As such they deal with highly sensitive financial information that will need to be encrypted. An awareness of cybersecurity measures, and protecting this information, is a key component of what they do.

### Work Environment

The largest employers of budget analysts include federal, state, and local governments and educational services; they are also often employed by professional, scientific, and technical services. Most budget analysts work full time, and overtime is sometimes required during development, mid-year, and final reviews of budgets. The pressures of deadlines and tight work schedules may be stressful.

Photo via iStock.com/MicroStockHub. [Used under license.]

## Occupation Interest

Budget analysts have strong mathematical skills and an interest in finances and meticulous work like recordkeeping. They are highly organized and patient, with expertise in conducting research, analyzing the data found and forecasting future trends based off said information.

### *Profile*

**Working Conditions:** Inside
**Education Needs:** Bachelor's Degree; Master's Degree
**Licensure/Certification:** Various
**Opportunities for Experience:** Budget- or finance-related work experience
**Interest Score:** CEI

## A Day in the Life—Duties and Responsibilities

Budget analysts advise organizations—including governments, private companies, and universities—about the details of their finances. They prepare annual and special reports and evaluate budget proposals. They analyze data to determine the costs and benefits of various programs, and they recommend funding levels based on their findings. Although government officials or top executives in a private company usually decide on an organization's budget, they rely on the work of budget analysts to prepare the information for that decision.

### *Duties and Responsibilities*

- Working with program and project managers to develop the organization's budget
- Reviewing managers' budget proposals and funding requests for completeness, accuracy, and compliance with laws and other regulations
- Combining program and department budgets into a consolidated organizational budget
- Explaining funding requests to others in the organization, to legislators, and to the public
- Helping top managers analyze proposed plans and finding alternatives if the projected results are unsatisfactory
- Monitoring organizational spending to ensure that it is within budget
- Informing program managers of the status and availability of funds
- Estimating future financial needs

Sometimes, budget analysts use cost-benefit analyses to review financial requests, assess program tradeoffs, and explore alternative funding methods. Budget analysts also may examine past budgets and research economic and financial developments that affect the organization's income and expenditures. Budget analysts may recommend cutting spending on particular programs or redistributing funds.

Throughout the year, budget analysts oversee spending to ensure that organizations comply with the budget and to determine whether certain programs need changes in funding. Analysts also evaluate programs to determine whether they are producing desired results.

In addition to providing technical analysis, budget analysts must communicate their recommendations effectively within the organization. For example, if there is a difference between the approved budget and actual spending, budget analysts may write a report explaining those discrepancies and recommend changes to reconcile them.

Budget analysts working in government may attend committee hearings to explain their recommendations to legislators. Occasionally, budget analysts evaluate how well a program is doing, assess policy, and draft budget-related legislation.

## WORK ENVIRONMENT

### Immediate Physical Environment

Most budget analysts work in offices, but some work from home. Budget analysts may travel to their clients' places of business in such cases where they may need to explain their findings in person.

### Human Environment

Budget analysts work with not just facts and figures but also people, and they will need to have a level of ease in social matters. Budget analysts must confer with program and project managers on developing budgets; as well, they may find themselves reviewing and explaining these budgets and budget proposals with others in the organization, legislators, and potentially the general public. As such they will be communicating with people in person, on video calls, on the phone, and by email.

### Technological Environment

Budget analysts use various software programs such as spreadsheets, as well as database and financial analysis software. Many organizations use Enterprise Resource Planning (ERP) programs as part of their budgeting process.

## EDUCATION AND TRAINING

### High School/Secondary

High school students interested in becoming budget analysts should focus on math, statistics, and business-related courses. Becoming a budget analyst requires at least a bachelor's degree, so students should prepare themselves for postsecondary education, including researching the schools they would like to consider attending to achieve that goal.

**Suggested High School Subjects**

- Algebra
- Biology
- Chemistry
- Civics
- Economics
- Geometry
- Pre-Calculus
- Statistics
- Trigonometry

**Related Career Pathways/Majors**

*Business Management & Administration Career Cluster*

- General Management Pathway
- Operations Management Pathway

*Finance Career Cluster*

- Accounting Pathway
- Business Finance Pathway

### Transferable Skills and Abilities

**Analytical Skills**

- Processing a variety of information
- Evaluating costs and benefits
- Solving complex problems

**Communication Skills**

- Explaining and defending their analyses and recommendations in meetings and legislative committee hearings

**Detail-oriented**

- Creating an efficient budget
- Carefully analyzing each budget item

**Math Skills**

- Analyzing, comparing, and interpreting facts and figures
- Using advanced math skills, such as calculus and statistical analysis
- Using financial-management software and programs

**Writing Skills**

- Presenting written technical information in a way that is understandable to the intended audience

## Postsecondary

Budget analysts typically need at least a bachelor's degree in fields such as business, social science, psychology, or mathematics. Because developing a budget requires numeracy and analytical skills, coursework in accounting, economics, and statistics is helpful.

Sometimes, budget- or finance-related work experience may be substituted for formal education.

**Related College Majors**

- Accounting
- Auditing
- Business Management & Administration
- Economics
- Financial Management
- General Business

## Fast Fact

**The cost of cybersecurity has increased significantly for businesses. In fact, according to research firm Gartner Inc., businesses will spend $1.75 trillion on cybersecurity over the following five years. By comparison, this branch of industry was "only" worth $3.5 billion in 2004. The COVID-19 pandemic and the tendency toward remote work are two factors that have contributed to this spike in cost.** Source: gartner.com

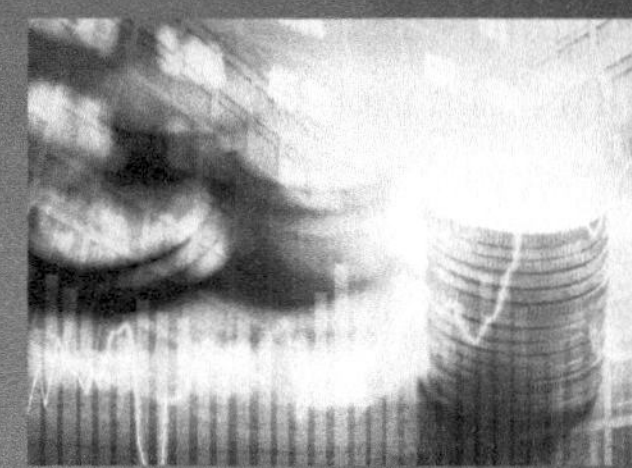

Photo via iStock.com/SARINYAPINNGAM. [Used under license.]

- Insurance
- International Business
- Marketing

### Adult Job Seekers

Adults seeking to become budget analysts should have a background or strength in math or finance. Schooling is still required, so thought should be given to an individual's long-term professional goals.

### Professional Certification and Licensure

Budget analysts working in federal, state or local government may earn the optional Certified Government Financial Manager (CGFM) credential from the Association of Government Accountants (AGA). CGFM candidates must have at least a bachelor's degree, abide by the AGA's Code of Ethics, pass examinations, and complete a designated period of professional-level experience in governmental financial management. To maintain certification, CGFMs must complete continuing education.

Although the CGFM is not required, having a designation may help with career advancement.

AICPA and the Chartered Institute of Management Accountants (CIMA) developed the Chartered Global Management Accountant (CGMA) designation as an internationally recognized professional credential. Candidates must complete a program, pass an exam, and meet a requirement for work experience.

## EARNINGS AND ADVANCEMENT

Earnings depend upon whether the individual is self-employed or works for a firm. The median annual wage for budget analysts was $79,940 in 2021. The lowest 10 percent earned less than $49,330, and the highest 10 percent earned more than $124,440.

Most budget analysts work full time, and overtime is sometimes required during development, mid-year, and final reviews of budgets.

## EMPLOYMENT AND OUTLOOK

Employment of budget analysts is projected to grow 3 percent from 2021 to 2031, slower than the average for all occupations. Within that ten-year timeframe, the number of budget analysts is predicted to jump from an estimated 50,400 persons to 51,800.

Despite limited employment growth, about 4,000 openings for budget analysts are projected each year, on average, over the decade. Most of those openings are expected to result from the need to replace workers who transfer to different occupations or exit the labor force, such as to retire.

Calls for efficient use of public funds will lead to continued demand for budget analysts to estimate program costs, develop budgets, and explain their findings to legislators and the public. Demand for these workers is somewhat tied to the government funding that is allocated for these positions. However, budget analysts manage resource allocation and will be needed even during times of tight budgets.

### Related Occupations

- Accountant/Auditor
- Actuary
- Cost Estimator
- Economist
- Financial Analyst
- Financial Examiner
- Financial Manager
- Insurance Underwriter
- Management Analyst
- Tax Examiner/Collector/Revenue Agent

## MORE INFORMATION

**American Accounting Association (AAA)**
9009 Town Center Parkway
Lakewood Ranch, FL 34202
941.921.7747
info@aaahq.org
aaahq.org
Forensic Accounting Division: aaahq.org/FIA

**American Association for Budget and Program Analysis (AABPA)**
P.O. Box 1157
Falls Church, VA 22041
AABPA@aabpa.org
www.aabpa.org

**American Board of Forensic Accounting (ABFA)**
1400 Easton Drive, Suite 137
Bakersfield, CA 93309
661.861.8861
staff@abfa.us

**Association of Certified Fraud Examiners (ACFA)**
Global Headquarters—The Gregor Building
716 West Avenue
Austin, TX 78701-2727
800.245.3321
www.acfe.com

**Association of Government Accountants (AGA)**
2208 Mount Vernon Avenue
Alexandria, VA 22301-1314
703.684.6931
www.agacgfm.org

**Association of International Certified Professional Accountants (AICPA)**
220 Leigh Farm Road
Durham, NC 27707-8110
888.777.7077
www.aicpa.org

**Forensic Accountants Society (FASNA)**
6200 Aurora Avenue, Suite 402W
Urbandale, IA 50322-2800
515.276.7000
www.fasna.org

**Forensic Certified Public Accountants Association (FCPAA)**
1120 Avenue of the Americas, Room 4057, 4th Floor
New York, NY 10036
626.521.9077
info@fcpachina.org
www.fcpai.org

**Forensic CPA Society**
P.O. Box 31060
Spokane, WA 99223
509.448.9318
www.fcpas.org

**Global Academy of Finance and Management (GAFM)**
P.O. Box 293
1670F E. Cheyenne Mountain Boulevard
Colorado Springs, CO 80906
www.gafm.com

**Government Finance Officers Association (GFOA)**
203 N. LaSalle Street, Suite 2700
Chicago, IL 60601
www.gfoa.org

**Information Systems Audit and Control Association (ISACA)**
1700 E. Golf Road, Suite 400
Schaumburg, IL 60173
847.253.1545
www.isaca.org

**Institute of Certified Forensic Accountants (ICFA)**
3422 Old Capitol Trail, Suite 1063
Wilmington, DE 19808-6192
302.425.5528
info@forensicglobal.org
www.forensicglobal.org

**Institute of Internal Auditors (IIA)**
1035 Greenwood Boulevard, Suite 401
Lake Mary, FL 32746
407.937.1111
na.theiia.org

**National Association of Forensic Accountants (NAFA)**
6635 West Commercial Boulevard, Suite 217
Tamarac, FL 33319
844.217.6975
mail@nafanet.com
www.nafanet.com

**National Association of State Budget Officers (NASBO)**
Hall of States Building
444 North Capitol Street, NW Suite 642
Washington, DC 20001
202.624.5382
nasbo-direct@nasbo.org
www.nasbo.org

---

*Daniel Vest*

## *Conversation With...*

# STEPHANIE GULLICKSON

Fiscal Management Analyst/Budget Analyst
North Dakota Office of Management and Budget, Bismarck, ND
Accounting field, 24 years

**What was your individual career path in terms of education/training, entry-level job, or other significant opportunity?**
I started working at the North Dakota Department of Public Instruction, which is the state K-12 education department, through a two-year high school paid internship program in 1990. After I graduated school, I applied for a full-time position there, and was hired as a secretary in 1992, a job I held for four years.

At the time, I thought I wanted to be a lawyer. I even earned my two-year paralegal degree from Interstate Business College in 1995. But I decided to change direction when, while working as a secretary, I had an opportunity to work and train with the staff accountant. I enjoyed the work she performed each day and realized that was what I really should be doing.

I was promoted to entry-level Account Technician in 1996 and worked under the staff accountant for many years until she retired. As opportunities to advance opened, I jumped at them to expand my experience. I was able to move from the entry-level account technician to the Director of Fiscal Management during my tenure at the Department of Public Instruction.

While working full-time, I received my bachelor's degree in Business Management and Accounting from University of Mary in Bismarck, ND in 2001, and went on to receive a Master of Public Administration Degree from University of North Dakota in Grand Forks in 2011.

In January 2016, after 25 years at the Department of Public Instruction, I was offered a Budget Analyst job with the North Dakota Office of Management and Budget, where I work now.

**What are the most important skills and/or qualities for someone in your profession?**
The most well-known skills are attention to detail, organizational skills, teamwork, critical thinking, time management and self-motivation. Honesty and integrity are important qualities because you handle the financial resources of state agencies or private sector businesses. You also need problem-solving skills and to think creatively. That might seem unusual for an accountant, but in my experience sometimes those very creative ideas truly make a difference! It's also important to pursue opportunities to increase soft skills such as interpersonal and

communication skills. At some point all those numbers we crunch as an accountant will need to be communicated and explained to people. These soft skills are especially important if you aspire to lead a team or be a manager.

**What do you wish you had known going into this profession?**
I wish I had known to take I.T.-related courses along with my accounting classes. Information technology skills are important because everything we do has a technology component, from the software that public accountants use to projects we help develop. A wider knowledge base makes you a more valuable, well-rounded employee, too.

**Are there many job opportunities in your profession? In what specific areas?**
Yes, almost 25 percent of government sector existing workforce is eligible to retire in the next five years. This large exit of public employees will open positions all over government in the accounting field. There are so many areas where accounting professionals are needed, from auditors, grant and contract managers, accounting budget specialists, accounting managers, business managers, financial institutions examiner, budget analysts all the way to the chief financial officer of the largest state agencies. Almost every state agency has a need for a professional accounting employee so even starting at an entry level position will give you valuable work experience.

**How do you see your profession changing in the next five years, how will technology impact that change, and what skills will be required?**
While the fundamentals of accounting probably won't change, the processes and methods we use do evolve. There are very few paper and pencil processes left and as things become even more paperless, it is important for us to adapt to these new methods. Specifically in public accounting, decisions on what to fund in a budget are based on data. An accountant who can understand the data and also calculate return on investment will have an advantage over someone who does not understand this process, so statistics courses will be important to you as a student! I also think in the next 5 years, public employees in general need to be adaptable to a variety of working situations, flexible working schedules and teleworking. Paperless offices, video teleconferencing and a variety of remote work tools have all become the norm.

**What do you enjoy most about your job? What do you enjoy least about your job?**
As a budget analyst, I enjoy being part of the team that builds the state budget. It's rewarding to start at the detailed level of each state agency and roll that up into a comprehensive statewide budget that is presented as the Governor's Executive Budget. We follow the executive budget throughout the legislative process and provide information as the statewide budget becomes final. It's satisfying to recognize that some of my budget recommendations survive the process and are part of the final product that will have an impact on the citizens in the state where I live.

The least favorite part of my job would probably be the same for most accountants, which is during an economic downturn when we need to make budget cuts. In any accounting position, it's very difficult to know that decisions you make can affect programs that may be discontinued or reduced, and which impact people's daily lives. But even during those times, accounting professionals add value by helping their agency staff look at the current processes and programs to find ways to become more efficient or discontinue things that do not make a difference in people's daily lives. In public accounting, each dollar that is saved or reduced is a taxpayer dollar.

**Can you suggest a valuable "try this" for students considering a career in your profession?**

Take advantage of any school-sponsored job shadowing opportunities, career fairs, or internships. This will give you a valuable look and conversation with accounting professionals already working in your chosen career path. An internship was how I found my full-time employment for the state of ND. Seek out a mentor who is already working in the field, be brave and ask these mentors for advice and help. They would love to share their knowledge with you! Keep in contact with your mentor so when a job opportunity opens up they will think of you and help guide you along the process of applying for and landing that job!

*This interview was originally published in 2020.*

## *Conversation With...*

# CLAIRE LODGE

Finance Consultant and Former Budget Analyst
Arlington, Virginia
Financial profession, 7 years
Budget analyst, 6 years

**What was your individual career path in terms of education/training, entry-level job, or other significant opportunity?**

I thought I wanted to do International Development/Development Economics. In fact, I have a B.A. in International Studies from American University and an M.S. in Development Economics from the University of Manchester in England. After graduation, however, I fell into budgeting. I had a job interview and the day-to-day tasks seemed analytical—which I liked from my economics background—but with a lot of people interaction. It seemed like a good fit for me.

To be a budget analyst, a specific degree is not needed. I'm good at math, but I think it's more important to be a logical thinker. The main thing, at least at the government agency where I worked, is writing ability and analytical skills. So, even though I went in without a budget-related educational background, I was hired and excelled.

We developed the budget by working within the department, then the Office of Management and Budget that is basically the White House budget wing, and, finally, with Congress.

**What are the most important skills and/or qualities for someone in your profession?**

Writing is really important. You have to be able to effectively write responses to inquiries from government departments and Congress and to write the agency's budget for Congress. But the other aspect of writing, which I think is sometimes taken for granted, is email writing. Tone is important. Sometimes you are giving bad news. So, even if it's only a four-sentence email, you have to be willing to reread it three times to make sure nothing can be misconstrued.

Analytical skills are critical. In the federal budgeting world, there is a mantra: Come to me with solutions, not problems. My boss actually wanted three solutions. You can't come up with solutions if you don't know how to analyze the problem.

You need to be organized. If you can't multi-task and keep track of your assignments, you will not excel in this career. Oftentimes, I was working on three different assignments and something time-pressing would come up. I had to know what I could afford to put to the back burner, what couldn't be left behind, and then communicate that to my bosses.

**What do you wish you had known going into this profession?**

I asked a lot of questions in my job interviews so I knew what to expect, at least as much as one can. The interviewee should ask questions to see if the job is a good fit.

**Are there many job opportunities in your profession? In what specific areas?**

Yes. The federal government needs budget analysts, as do state and local governments.

**How do you see your profession changing in the next five years, what role will technology play in those changes, and what skills will be required?**

I think that there will always be budget analysts in the public sector. Technology has made their jobs easier. I have no idea how budget analysts were able to do what they do before computers, and really before Excel.

**What do you like most about your job? What do you like least about your job?**

Being a budget analyst is not a routine job. I might do something different every day. Also, there may be crunch times, especially in public sector budgeting where you can plan on working later than normal at times like the end of a fiscal year. The lack of daily routine was a negative only because I couldn't make plans for the same time every day.

**Can you suggest a valuable "try this" for students considering a career in your profession?**

A lot of budget analysts learn on the job. The closest "try this" would be budgeting in your own life, but that leaves out some of the big parts of the federal budgeting process such as questions from Congress, submitting your budget for approval, and working with different entities.

*This interview was originally published in 2015.*

# Claims Adjuster/ Appraiser/Examiner/ Investigator

### Snapshot

**Career Cluster(s):** Business Management & Administration; Finance; Human Services

**Interests:** Accounting; Finance; Data; Statistics; Recordkeeping; Insurance; Investigating

**Earnings (Yearly Average):** $64,710

**Employment & Outlook:** Decline in Growth Expected

## OVERVIEW

### Sphere of Work

Claims adjusters, appraisers, examiners, and investigators evaluate insurance claims. They decide whether an insurance company must pay a claim and if so, how much. In a business landscape grown increasingly technological, part of this means uncovering damages resulting from real and alleged cyberattacks.

Photo via iStock.com/AndreyPopov. [Used under license.]

### Work Environment

Claims adjusters and examiners work in offices when reviewing documents and conducting research. They work outside when examining damaged property. Appraisers and investigators work outside more often, inspecting damaged automobiles and buildings and conducting surveillance. Auto damage appraisers spend much of their time at automotive body shops estimating vehicle damage costs.

### *Profile*

**Working Conditions:** Mainly Inside
**Education Needs:** High School or equivalent; Bachelor's Degree sometimes required
**Licensure/Certification:** Various
**Opportunities for Experience:** On-the-Job Training
**Interest Score:** CEI

Workers who inspect damaged buildings must be careful around potential hazards, such as collapsed roofs and floors, as well as weakened structures.

## Occupation Interest

Claims adjusters have an interest in conducting research and using their analytical skills to reach conclusions about the work sites they experience. They will have an interest as well in dealing with people, reading verbal and non-verbal cues, and may have an interest in working an unconventional sort of job with irregular hours.

## A Day in the Life—Duties and Responsibilities

Claims adjusters, appraisers, examiners, and investigators have varying duties, depending on the type of insurance company they work for. They must know a lot about what their company insures. For example, workers in property and casualty insurance must know housing and construction costs so that they can properly evaluate damage from floods or fires. Workers in health insurance must be able to determine which types of treatments are medically necessary and which are questionable.

### *Duties and Responsibilities*

- Investigating, evaluating, and settling insurance claims
- Determining whether the insurance policy covers the loss claimed
- Deciding the appropriate amount the insurance company should pay
- Ensuring that claims are not fraudulent
- Contacting claimants' doctors or employers to get additional information on questionable claims
- Talking with legal counsel about claims when needed
- Negotiating settlements
- Authorizing payments

Adjusters inspect property damage or personal injury claims to determine how much the insurance company should pay for the loss. They might inspect a home, a business, or an automobile.

Adjusters interview the claimant and witnesses, inspect the property, and do additional research, such as look at police reports. They may consult with other workers, such as accountants, architects, construction workers, engineers, lawyers, and physicians, who can offer a more expert evaluation of a claim.

Adjusters gather information—including photographs and statements, either written or recorded on audio or video—and put together a report for claims examiners to review. When the examiner approves the claim, the adjuster negotiates with the policyholder and settles the claim.

If the claimant contests the outcome of the claim or the settlement, adjusters work with attorneys and expert witnesses to defend the insurer's position.

Some claims adjusters work as public adjusters. Often, they are hired by claimants who prefer not to rely on the insurance company's adjuster. The goal of adjusters working for insurance companies is to save as much money for the company as possible. The goal of a public adjuster working for a claimant is to get the highest possible amount paid to the claimant. They are paid a percentage of the settled claim.

Appraisers estimate the cost or value of an insured item. Most appraisers who work for insurance companies and independent adjusting firms are auto damage appraisers. They inspect damaged vehicles after an accident and estimate the cost of repairs. This information then goes to the adjuster, who puts the estimated cost of repairs into the settlement.

Claims examiners review claims after they are submitted to make sure claimants and adjusters followed proper guidelines. They may help adjusters with complicated claims or when, for example, a natural disaster occurs and the volume of claims increases.

Examiners who work for health insurance companies review health-related claims to see whether the costs are reasonable, given the diagnosis. After they review the claim, they authorize appropriate payment, deny the claim, or refer the claim to an investigator.

Examiners who work for life insurance companies review the causes of death and pay particular attention to accidents, because most life insurance companies pay additional benefits if a death is accidental. Examiners also may review new applications for life insurance policies to make sure that the applicants have no serious illnesses that would make them a high risk to insure.

Insurance investigators handle claims in which the company suspects fraudulent or criminal activity such as arson, staged accidents, or unnecessary medical treatments. The severity of insurance fraud cases varies, from overstated claims of vehicle damage to complicated fraud rings. Investigators often do surveillance work. For example, in the case of a fraudulent workers' compensation claim, an investigator may covertly watch the claimant to see if he or she does anything that would be suspicious based on injuries stated in the claim.

## WORK ENVIRONMENT

### Immediate Physical Environment

Claims adjusters and examiners work both indoors and outdoors. A majority of their time may be spent in-office, reviewing documents and conducting research. This is especially true if their work specifically deals with the world of cybersecurity and investigating cybercrime, where most if not all of the evidence they examine is "virtual" in nature. Appraisers and investigators work outside more often, inspecting damaged property such as automobiles and buildings, as well as conducting surveillance. In the case of those who do work around damaged buildings, these appraisers must be careful around potential hazards.

### Human Environment

Claims adjusters and investigators may work in teams and resultingly come into contact with others on a regular basis. Strong communication skills will be necessary, as they need to interpret the data observed and provided and convey their findings in a clear fashion when dealing with clients.

### Technological Environment

Claims adjusters, appraisers, examiners and investigators use technology such as various software programs including Enterprise resource planning (ERP) software, Financial Analysis software, and Computer aided design (CAD) software, and more.

## EDUCATION AND TRAINING

### High School/Secondary

A high school diploma or equivalent is typically required for a person to work as an entry-level claims adjuster or examiner. High school students interested in becoming claims adjusters or examiners should focus on math, statistics, and business-related courses. Those entering this profession should know that can employers can sometimes require a bachelor's degree, so students should prepare themselves for possible postsecondary education, including researching the schools they would like to consider attending to achieve their goals.

**Suggested High School Subjects**

- Algebra
- Civics
- Economics
- English
- Pre-Calculus
- Psychology
- Statistics

**Related Career Pathways/Majors**

*Finance Career Cluster*

- Accounting Pathway
- Insurance Pathway

*Human Services Career Cluster*

- Consumer Services Pathway

## Transferable Skills and Abilities

**Analytical Skills**

- Evaluating whether the insurance company is obligated to pay a claim determine the amount to pay
- Considering various pieces of information to reach a decision

**Communication Skills**

- Getting information from a range of people, including claimants, witnesses, and medical experts
- Knowing the right questions to ask in order to gather the information needed

**Detail-oriented**

- Carefully reviewing documents and damaged property, because small details can have large financial consequences

**Interpersonal Skills**

- Meeting with claimants and others who may be upset by the situation that requires a claim or by the settlement the company is offering
- Being understanding, yet firm regarding company policy

### Postsecondary

Postsecondary education is not a strict requirement: it depends on the employer whether or not a high school diploma will suffice. Still, a postsecondary education in finance, etc., can certainly open more doors to future opportunities and it is ultimately up to the individual.

Auto damage appraisers typically have either a postsecondary nondegree award or experience working in an auto repair shop, identifying and estimating the cost of automotive repair. Many vocational schools and some community colleges offer programs in autobody repair that teach students how to estimate the cost of repairing damaged vehicles.

Entry-level claims adjusters, examiners, and investigators work on small claims under the supervision of an experienced worker. As they learn more about claims investigation and settlement, they are assigned larger, more complex claims.

Auto damage appraisers typically get on-the-job training, which may last several months. This training usually involves working under the supervision of an experienced appraiser while estimating damage costs, until the employer decides that the trainee is ready to do estimates on their own.

**Related College Majors**

- Accounting
- Auditing
- Business Management & Administration
- Economics
- Entrepreneurial Studies
- Financial Management
- General Business
- Insurance
- International Business

## Adult Job Seekers

Adults seeking to become claims adjusters, appraisers, examiners or investigators, will typically need to have at least a high-school diploma to get an entry-level position in one of these jobs. However, many employers prefer a bachelor's degree, or equivalent work experience in insurance. For investigator jobs, people trained as law enforcement officers or private investigators are considered great prospects as they have good interviewing and interrogation skills.

## Professional Certification and Licensure

Licensing requirements for claims adjusters, appraisers, examiners, and investigators vary by state. Some states have few requirements; others require either completing prelicensing education or receiving a satisfactory score on a licensing exam (or both). Jobseekers should verify the licensing laws with the state and locality in which they want to work.

In some states, claims adjusters employed by insurance companies do not have to become licensed themselves because they can work under the company license.

Public adjusters may need to meet separate or additional requirements.

Some states that require licensing also require a certain number of continuing education credits per year to renew the license. Federal and state laws and the outcomes of claim disputes adjudicated in court affect how the claims must be handled and what insurance policies can and must cover. Examiners working on life and health claims must stay up to date on new medical procedures and prescription drugs. Examiners working on auto claims must be familiar with the most recent car models and repair techniques. To fulfill their continuing education requirements, workers can attend classes or workshops, write articles for claims publications, or give lectures and presentations.

The National Insurance Producer Registry (NIPR) provides information about state licensing requirements. (See More Information section below)

## EARNINGS AND ADVANCEMENT

The median annual wage for claims adjusters, examiners, and investigators was $65,080 in 2021. The lowest 10 percent earned less than $41,490, and the highest 10 percent earned more than $101,290.

The median annual wage for insurance appraisers, auto damage was $62,680 in May 2021. The lowest 10 percent earned less than $47,240, and the highest 10 percent earned more than $91,680.

Most claims adjusters, appraisers, examiners, and investigators work full time. However, their work schedules may vary.

Adjusters often arrange their work schedules to accommodate evening and weekend appointments with clients. This requirement sometimes results in adjusters working irregular schedules, especially when they have a lot of claims to review.

Insurance investigators often work irregular schedules because of the need to contact people who are not available during normal business hours. Early morning, evening, and weekend work is common.

In contrast, auto damage appraisers typically work regular hours and rarely work on weekends.

## EMPLOYMENT AND OUTLOOK

Overall employment of claims adjusters, appraisers, examiners, and investigators is projected to decline 6 percent from 2021 to 2031.

Despite declining employment, about 23,200 openings for claims adjusters, appraisers, examiners, and investigators are projected each year, on average, over the decade. All of those openings are expected to result from the need to replace workers who transfer to other occupations or exit the labor force, such as to retire.

## Fast Fact

**In their 2022 report on global ransomware damage costs, Cybersecurity Ventures predicted that ransomware attacks would target businesses every 11 seconds. This estimate was a 21% increase from the previous forecast of every 14 seconds, that was estimated in 2019. Claims professionals will therefore need possess wide knowledge in order to help victims impacted by an attack, such as the difference between a phishing attack, data compromise, wire fraud, password attack or other cybercrime.** Source: propertycasualty360.com

Photo via iStock.com/Black_Kira. [Used under license.]

Technology is expected to automate some of the tasks that these workers currently perform. For example, computer software can evaluate photographs of damaged property and calculate an estimated claim amount. In addition, data collection and processing speed will continue to increase, which will improve efficiency and make workers more productive.

The number of natural disasters, such as floods and fires, influences demand for claims adjusters in property and casualty insurance. Future increases in the number of natural disasters could result in some employment growth for claims adjusters in the field.

### Related Occupations

- Automotive Body/Glass Repairer
- Automotive Service Technician/Mechanic
- Construction/Building Inspector
- Cost Estimator
- Fire Inspector
- Property Appraiser/Assessor

## MORE INFORMATION

**The Institutes**
720 Providence Road, Suite 100
Malvern, PA 19355
800.644.2101
customersuccess@theinstitutes.org
www.theinstitutes.org

**International Claim Association (ICA)**
1800 M Street, NW, 400 South
Washington, DC 20036
202.452.0143
memberservices@claim.org
www.claim.org

**National Association of Public Insurance Adjusters (NAPIA)**
1400 Village Square Blvd, Suite 3—187
Tallahassee, FL 32312
703.433.9217
info@napia.com
napia.com

**National Insurance Producer Registry (NIPR)**
1100 Walnut Street, Suite 1500
Kansas City, MO 64106
www.nipr.com

---

*Daniel Vest*

## *Conversation With...*
## MARY ANNE MEDINA

Director of Business Development & Training
One Call Claims, Orlando, Florida
Insurance adjuster, 25+ years

**What was your individual career path in terms of education/training, entry-level job, or other significant opportunity?**

After I graduated from college, I taught for a year. Several people in my family worked in claims. It was during a time when the industry was under pressure to hire women. I went to work for a large adjusting firm that had its own training facility, so I got a lot of training. It intrigued me to be able to move into a position with so many benefits and a company vehicle and room for promotions and the ability to learn many aspects of the industry: property claims, catastrophe claims, casualty, litigation. I eventually moved to upper management.

Prior to the job that I have now, I was vice president of Vale Training Solutions, which is a training facility for adjusters. My career path was to go out and experience everything in the field, then bring it back and train others. It came full circle.

**What are the most important skills and/or qualities for someone in your profession?**

I think communication is probably the priority. If you're a catastrophe adjuster, you're dealing with someone who's been devastated and lost their home. You need to be able to guide them through the process. Or if you're called to testify, you have to be able to communicate in a courtroom. You have to be able to communicate with different cultural groups. We constantly get calls for what we call "soft skills." The new people coming in aren't used to not sending that instant text and their writing skills to communicate and explain are not there.

And secondly, being organized. Being an adjuster, especially a catastrophe adjuster, can be very demanding— trying to see everyone and return every call and every email and get people back on their feet.

You have to genuinely like people. If you can't deal with confrontation, it's not for you. You have to be empathetic. It may seem like a mild fender bender to you, but to this grandma who's never had an accident in her life, it's major.

**What do you wish you had known going into this profession?**

I wish I had known how physically demanding the job can be. When it's 105 degrees and you're putting a 70-pound ladder up on a roof 10 times a day or you're going through a commercial building that's been burned and you can't breathe because there's soot in your nasal passages, it's tough. When you go to a catastrophe, you're hoping you can find a hotel room; there may be no water or power. And it's emotionally demanding. I've seen things that I'll never forget. I had a claim for a family in Miami where during the height of a storm—and of course, they blamed themselves, they didn't board the windows up—well, a piece of the window came through and hit their 9-year-old across the chest and he bled to death. Things like that really do impact you.

**Are there many job opportunities in your profession? In what specific areas?**

Yes. No matter what happens with the economy, there are still tornadoes and storms and car accidents and people falling down in grocery stores.

**How do you see your profession changing in the next five years? What role will technology play in those changes, and what skills will be required?**

Technology constantly changes and you have to keep up. Drones can fly over and take pictures for adjusting purposes. We already have satellite capability in areas where the roof is too steep or you can't get to the area. You can examine damage, get in your car, dial your cell phone and update a voice-activated diary: 'Went to Mrs. Jones home, she had 6 feet of water' and Mrs. Jones can log into her claim and see what's going on.

Everything is fast, fast, fast, but now customers are saying, "I got this estimate and I don't understand it." I really see a shift in realizing that fast is not always better, that you can't replace human contact.

**What do you enjoy most about your job? What do you enjoy least about your job?**

Meeting people and helping people. It's a tough industry. I've always looked at it as really helping people get their lives back together.

For a while, what I liked least was the travel. I was missing a lot of the joys of life: my kids, my husband. Catastrophe adjusters can be gone for long periods and you can't get that time back.

**Can you suggest a valuable "try this" for students considering a career in your profession?**

There are claims offices located pretty much everywhere. Go in and talk to an adjuster. There's so much out there that you can read, like *Claims Magazine* and *Claims Management Magazine*.

*This interview was originally published in 2016.*

# Computer and Information Research Scientist

### Snapshot

**Career Cluster(s):** Information Technology; Science, Technology, Engineering & Mathematics

**Interests:** Computers; Data Science; Technology

**Earnings (Yearly Average):** $131,490

**Employment & Outlook:** Faster Than Average Growth Expected

## OVERVIEW

### Sphere of Work

Computer and information research scientists design innovative uses for new and existing technology. They study and solve complex problems in computing for business, science, medicine, and other fields. A key problem these research scientists face is the need to develop new and innovative cybersecurity measures to fend off attacks from cybercriminals whose methods are always evolving.

Photo via iStock.com/Georgijevic. [Used under license.]

### Work Environment

Some scientists collaborate with engineers or other specialists or research scientists in different locations and do much of their work online.

## Occupation Interest

Computer and information research scientists are highly educated, disciplined thinkers, whose work involves a great deal of analytical thought. Searching for facts and figuring out problems mentally, they will need to follow a set of procedures and routines, and having done this foundational work, they will be free to let their creative side take over. Through outside-of-the-box thinking and experimentation, they can find new solutions to the design, testing, and improving of computer systems.

### Profile

**Working Conditions:** Inside
**Physical Strength:** Light Work
**Education Needs:** Bachelor's Degree; Masters; PhD
**Licensure/Certification:** Not Required
**Opportunities for Experience:** On-the-Job Training
**Interest Score:** IRC

## A Day in the Life—Duties and Responsibilities

Computer and information research scientists create and improve computer software and hardware.

To create and improve software, computer and information research scientists work with algorithms: sets of instructions that tell a computer what to do. Some difficult computing tasks require complex algorithms, which these scientists simplify to make computer systems as efficient as possible. These simplified algorithms may lead to advancements in many types of technology, such as machine learning systems and cloud computing.

To improve computer hardware, these scientists design computer architecture. Their work may result in increased efficiencies, such as better networking technology, faster computing speeds, and improved information security.

### Duties and Responsibilities

- Exploring problems in computing and developing theories and models to address those problems
- Collaborating with scientist and engineers to solve complex computing problems
- Determining computing needs and system requirements
- Developing new computing languages, software systems, and other tools to improve how people work with computers
- Designing and conducting experiments to test the operation of software systems, frequently using techniques from data science and machine learning
- Analyzing the results of their experiments
- Writing papers for publication and presenting research finding at conferences

## OCCUPATION SPECIALTIES

### Programming

Some computer and information research scientists study and design new programming languages that are used to write software. New languages make software writing efficient by improving an existing language, such as Java, or by simplifying a specific aspect of programming, such as image processing.

### Robotics

These scientists study the development and application of robots. They explore how a machine can interact with the physical world. For example, they may create systems that control the robots or design robots to have features such as information processing or sensory feedback. Artificial intelligence (AI) increasingly plays a role in this developing industry.

Some computer and information research scientists work on multidisciplinary projects with electrical engineers, computer hardware engineers, and other specialists. For example, robotics specialists and engineers who design robots' hardware may team up to test whether the robots complete tasks as intended.

## WORK ENVIRONMENT

### Immediate Physical Environment

Work settings for computer and information research scientists could be anywhere with a solid internet connection and include Federal office buildings; college and university research buildings; medical and scientific laboratories; perhaps even a home office, as remote work grows more and more prevalent.

### Human Environment

While it can be cerebral and sometimes solitary work, some research scientists find themselves in collaboration with engineers or other specialists in order to solve complex computing problems. Computer and information research scientists also frequently write papers for publication and then present their findings in front of their peers and colleagues at various conferences.

### Technological Environment

A variety of computer software skills and know-how are necessarily part of the job, as computer and information research scientists will be designing, testing and improving computer systems continuously over the course of their careers.

Analytical software (minitab), Development software (Apache Kafka, Eclipse IDE), Operating systems (Bash, Oracle Solaris, UNIX) among many other varieties of software may be valuable and even necessary.

## EDUCATION AND TRAINING

### High School/Secondary

High school computer classes can begin to lay the foundations for prospective computer and information research scientists. The basic computing and research skills anyone may need should begin here. As well, a strong grasp of mathematics is a must; high school math and science classes must be pursued.

**Suggested High School Subjects**

- Algebra
- Computer Science
- Earth or Life or Physical Science
- English
- Geometry
- Physics
- Trigonometry

**Related Career Pathways/Majors**

*Information Technology Career Cluster*

- Programming & Software Development Pathway

*Science, Technology, Engineering & Mathematics Career Cluster*

- Engineering & Technology Pathway
- Science & Mathematics Pathway

**Transferable Skills and Abilities**

**Analytical Skills**
- Evaluating the results of research

**Communication Skills**
- Clearly explaining research, including to a nontechnical audience
- Writing papers for publication
- Presenting their research at conferences

**Detail oriented**
- Paying close attention to their work, such as when testing the systems they design

**Interpersonal Skills**
- Working effectively with programmers and managers
- Working on teams with engineers or other specialists

**Logical thinking**
- Using sound reasoning when working on algorithms

**Math Skills**
- Possessing a solid grasp of advanced math and other technical subjects critical to computing

**Problem-solving Skills**
- Thinking creatively to find innovative solutions in their research

### Postsecondary

Computer and information research scientists typically need a master's or higher degree in computer science or a related field, such as computer engineering. A master's degree usually requires 2 to 3 years of study after earning a bachelor's degree in a computer-related field, such as computer science or information systems. Some employers prefer to hire candidates who have a PhD. Others, such as the federal government, may hire candidates who have a bachelor's degree in computer and information technology.

Computer and information research scientists who work in a specialized field may need knowledge of that field. For example, those working on biomedical applications may need to have studied biology.

**Related College Majors**

- Computer Engineering
- Computer Science
- Cybersecurity
- Information Science
- Mathematics
- Software Engineering
- Systems & Network Administration

## EARNINGS AND ADVANCEMENT

The median annual wage for computer and information research scientists was $131,490 in 2021. The lowest 10 percent earned less than $74,210, and the highest 10 percent earned more than $208,000.

Most computer and information research scientists work full time. Some computer and information research scientists advance to become computer and information systems managers.

## EMPLOYMENT AND OUTLOOK

Computer and information research scientists held about 33,500 jobs in 2021. Employment is projected to grow 22 percent from 2021 to 2031, much faster than the average for all occupations.

About 3,200 openings for computer and information research scientists are projected each year, on average, over the decade. Many of those openings are expected to result from the need to replace workers who transfer to different occupations or exit the labor force, such as to retire.

## Famous First

**The field of cybersecurity research started as a grassroots effort through the Phreaker movement. Phreaking was a precursor of hacking, in which technically skilled people illegitimately manipulated telecommunications systems for various purposes (like making free long-distance calls). Phreakers would reverse-engineer hardware and analog communication protocols to learn how the telephone systems worked and operated. Pictured is a phreaking device known as a Blue Box, designed and built by Steve Wozniak and sold by Steve Jobs before they founded Apple, now on display at the Computer History Museum.** Source: sciencedirect.com

Photo by Maksym Kozlenko, via Wikimedia Commons.

The research and development conducted by computer and information research scientists turn ideas into technology. As demand for new and better technology grows, demand for computer and information research scientists will grow as well.

Rapid growth in data collection by businesses will lead to an increased need for data-mining services. Computer and information research scientists will be needed to write algorithms that help businesses make sense of very large amounts of data.

A growing emphasis on cybersecurity also should lead to new jobs because computer and information research scientists will be needed to find innovative ways to prevent potential cyberattacks. In addition, an increase in demand for software may increase the need for computer and information research scientists who create new programming languages to make software writing more efficient.

### Related Occupations

- Computer/Information Systems Manager
- Computer Hardware Engineer
- Computer Network Architect
- Computer Programmer
- Computer Systems Analyst
- Database Administrator/Architect

- Information Security Analyst
- Network/Computer Systems Administrator
- Software Developer
- Web Developer/Digital Designer

## MORE INFORMATION

**Association for Computing Machinery (ACM)**
1601 Broadway, 10th Floor
New York, NY 10019-7434
212.869.7440
acmhelp@acm.org
www.acm.org

**Association for Women in Computing (AWC)**
P.O. Box 2768
Oakland, CA 94602
info@awc-hq.org
www.awc-hq.org

**Computing Research Association (CRA)**
1828 L Street, NW, Suite 800
Washington, DC 20036
202.234.2111
cra.org

**IEEE Computer Society**
2001 L Street NW, Suite 700
Washington, DC 20036
202.371.0101
help@computer.orrg
computer.org

**International Association of Computer Science and Information Technology (IACSIT)**
contact@iacsit.org
www.iacsit.org

**National Center for Women & Information Technology (NCWIT)**
1909 26th Street, 2nd Floor
Boulder, CO 80302
303.735.6671
info@ncwit.org
ncwit.org

**USA JOBS**
1900 E Street NW
Washington, DC 20415
www.usajobs.gov

---

*Daniel Vest*

## *Conversation With...*
## WILLIAM P. SMYTH

IT Program Analyst, Federal Government
Washington, D.C.
IT Program Analyst, 31 years

**What was your individual career path in terms of education/training, entry-level job, or other significant opportunity?**

I went to Hampton University in Virginia for electrical engineering, but dropped out and joined the U.S. Marine Corps. I ended up doing aviation electronics, which I studied at a community college. I spent six years in the Marines, and played football for the Marine Corps team. NFL teams were scouting me. I went to Atlanta, but wasn't what they were looking for. Then I came to D.C. because the Redskins wanted to see me. However, a motorcycle accident ended my NFL dreams at age 26.

After I healed, I started looking for jobs. At the time, computers were not widespread. I got a downtown D.C. territory repairing electronic typewriters. The law firm Steptoe & Johnson was a client and they had desktop computers. I'd never seen one before, but somebody saw a problem with a cable and asked me to look at it. I traced the cable back, plugged it in, and said, "Try now." That prompted me to get into computers, and I never looked back.

I spent many overnights at law firms setting up networks with little support. It made me really good at my job, even though it could be frustrating. But I'm the kind of person who is persistent and likes to troubleshoot.

Eventually, I moved on to a company called Banctech that built banking machines and computers. I did phone support on networks and was there about five years. The company moved to Texas; I didn't want to go to Texas. So, I found a job as a contractor for a federal agency during Bill Clinton's first term. The agency had no networks or desktop computers, so I was the first one to put in their networks.

After 15 years as a contractor, I was hired eight years ago as a federal employee to support a departmental network and help desk, where I work with a team of 25 people. We handle Freedom of Information requests and run a certification and accreditation program. If someone internally wants to set up a website—for instance, so people can get information from the DOE about solar panels—we review the site and run tests to look for security weaknesses. This is an annual process. In addition, I get pulled in to run projects that are highly technical.

For instance, you can't just plug into the internet from our system. Vendors and others who need to do that come to me to find out how.

**What are the most important skills and/or qualities for someone in your profession?**
You really have to understand the theory of computers. If you understand how everything works from the bottom up, you'll be able to troubleshoot. I have to stress that you need a love for this, because things change every six months. If you're not willing to be constantly updated, you'll be far behind pretty quickly.

**What do you wish you had known going into this profession?**
I wish I had known that hacking would become such a big deal. I was on the leading edge when I started, when the internet was DOS-based. I could go onto important websites and there was no protection. Looking back, I could have come up with something to fix vulnerabilities and I wouldn't be working now!

**Are there many job opportunities in your profession? In what specific areas?**
Absolutely. Every day in the news, somebody is getting hacked. However, the security field is intense. You're on the clock 24 hours. Not everybody wants to do that. You can make a lot of money in the security field if you can get those difficult Cisco security certifications. But you can get dismissed real quick, too, if somebody gets into a system you're watching.

**How do you see your profession changing in the next five years, what role will technology play in those changes, and what skills will be required?**
I think it's going to change based on protocols. New protocols—computer languages—are being developed all the time, because every company is trying to stop hackers. And hackers are constantly working on how to get in. You really need to learn programming and protocols.

**What do you enjoy most about your job? What do you enjoy least about your job?**
I most enjoy troubleshooting. When an organization has a problem, I get to figure it out. I least like supporting difficult fellow employees who work in departments other than IT. For instance, I have to approve purchases. If somebody wants a 27-inch monitor, they need to tell me why—and not just assume that they will get it.

**Can you suggest a valuable "try this" for students considering a career in your profession?**
On websites such as http://etherealmind.com or https://learningnetwork.cisco.com/community/learning_center/games, you can play security games, take security tests, or join groups to discuss the topic. You'll find out if you have a real interest in security.

*This interview was originally published in 2019.*

# Computer Network Architect

## Snapshot

**Career Cluster(s):** Information Technology

**Interests:** Computing; Data; Design; Science & Technology

**Earnings (Yearly Average):** $120,520

**Employment & Outlook:** As Fast As Average Growth Expected

## OVERVIEW

### Sphere of Work

Computer network architects design and build data communication networks, including local area networks (LANs), wide area networks (WANs), and Intranets. These networks range from small connections between two offices to next-generation networking capabilities such as a cloud infrastructure that serves multiple customers. Network architects must have extensive knowledge of an organization's business plan to design a network that can help the organization achieve its goals. By the very nature of their work, matters of cybersecurity are at top of mind for any computer network architect, and their jobs will frequently put them in touch with security professionals so as to offer their clients the most secure networks possible.

Photo via iStock.com/gorodenkoff. [Used under license.]

### Work Environment

Computer network architects spend most of their time in offices, but occasionally work in server rooms where they have access to the hardware that make up an organization's computer and information network.

## Profile

**Working Conditions:** Inside
**Education Needs:** Bachelor's Degree
**Licensure/Certification:** Varies by company and software type
**Opportunities for Experience:** Related occupations
**Interest Score:** ICE

### Occupation Interest

Computer network architects have a strong interest in technology and design, and possess a fascination for how things work. They are highly organized thinkers, meticulous and patient. Individuals with an aptitude for painstaking and thorough mental work, they will conduct research, provide analysis and regularly confer with others in a team setting to build better networks.

### A Day in the Life—Duties and Responsibilities

Computer network architects, or network engineers, design and deploy computer and information networks. After deployment, they also may manage the networks and troubleshoot any issues as they arise. Network architects also predict future network needs by analyzing current data traffic and estimating how growth will affect the network. With all network systems, cybersecurity is of the utmost importance. Network architects therefore often work with security professionals (i.e. information security officers), creating password and system access protocols and establishing safeguards against cybersecurity threats, including malware, spyware, DoS (denial of service), and other forms of security breach.

## Duties and Responsibilities

- Creating plans and layouts for data communication networks
- Presenting plans to management and explaining why they are in the organization's best interest to pursue them
- Considering information security when designing networks
- Upgrading hardware, such as routers or adaptors, and software, such as network drivers, as needed to support computer networks
- Researching new networking technologies to determine what would best support their organization in the future

## WORK ENVIRONMENT

### Immediate Physical Environment

Most computer network architects work in offices, but some may work occasionally from home. Frequently they work with large firms, in coordination with Chief Technology Officers, and have teams of computer hardware engineers working under them. Computer network architects will sometimes have to commute to office environments in order to present information to management.

### Human Environment

Computer network architects must work with different types of employees to successfully design and implement computer and information networks; sometimes they are put in leadership positions where they have to communicate directions to a team of engineers. Computer network architects will sometimes work with other IT workers, such as network and computer system administrators and computer and information systems managers to ensure workers' and clients' networking needs are being met. They also must work with equipment and software vendors to manage upgrades and support the networks.

### Technological Environment

The technological needs of the job are many: computer network architects must possess an extensive knowledge of various types of operational, security and wireless systems. A level of mastery over various types of software programs, programming skills, networking skills, firewall and security skills, and more, is essential.

## EDUCATION AND TRAINING

### High School/Secondary

High school students interested in becoming computer network architects should focus on math, statistics, and science-related courses. Becoming one requires at least a bachelor's degree, so students should prepare themselves for postsecondary education, including researching the schools they would like to consider attending to achieve that goal.

**Suggested High School Subjects**

- Algebra
- Biology
- Chemistry
- Civics
- Earth or Life or Physical Science
- Economics
- English
- Geometry

- Physics
- Pre-Calculus
- Psychology
- Statistics
- Trigonometry

**Related Career Pathways/Majors**

*Business Management & Administration Career Cluster*

- Business Information Management Pathway
- Operations Management Pathway

*Information Technology Career Cluster*

- Network Systems Pathway
- Programming & Software Development Pathway

## Transferable Skills and Abilities

**Analytical Skills**

- Examining data networks and deciding how to best connect the networks based on the needs and resources of the organization

**Detail-oriented**

- Creating comprehensive plans of the networks they are creating with precise information describing how the network parts will work together

**Interpersonal Skills**

- Working with different types of employees to successfully design and implement computer and information networks

**Leadership Skills**

- Directing teams of engineers, such as computer hardware engineers, who build the networks they have designed

**Organizational Skills**

- Coordinating many different types of communication networks and make sure they work well together

## Postsecondary

Computer network architects typically need at least a bachelor's degree in computer and information technology, engineering, or a related field. Degree programs in a computer-related field give prospective network architects hands-on experience in classes such as network security or database design. These programs prepare network architects to be able to work with the wide array of technologies used in networks.

Employers of network architects sometimes prefer applicants to have a master's of business administration (MBA) in information systems. MBA programs generally require 2 years of study beyond the undergraduate level and include both business and computer-related courses.

Network architects generally need to have at least 5 to 10 years of experience working with information technology (IT) systems. They often have experience as a network and computer system administrator but also may come from other computer-related occupations.

**Related College Majors**

- Business Management & Administration
- Computer Science
- Financial Management

## Famous First

**Established in the late 1960s, ARPANET (Advanced Research Projects Agency Network) was an experimental computer network and a forerunner of the internet. ARPANET was a U.S. Defense Department initiative that linked computers at Pentagon research institutions via telephone lines.**

Source: Britannica.com

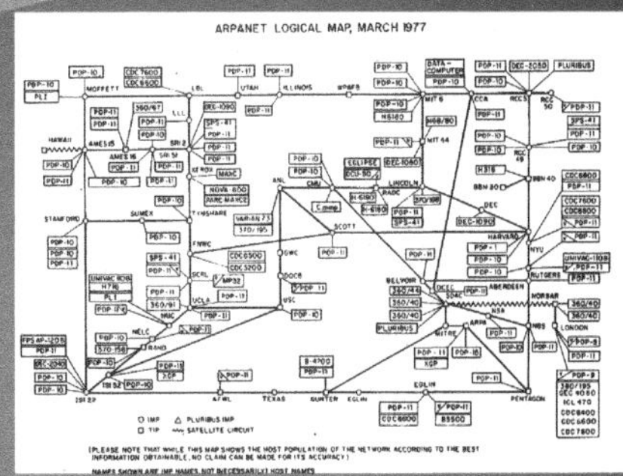


Photo via Wikimedia Commons. [Public domain.]

- General Business
- International Business
- Marketing

### Adult Job Seekers

Adults seeking to become computer network architects will typically need a bachelor's degree and experience in a related field, such as database administrator or computer systems analyst.

### Professional Certification and Licensure

Certification programs are generally offered by product vendors or software firms. Vendor-specific certification verifies a set of skills to ensure network architects are able to work in specific networking environments. Companies may require their network architects to be certified in the products they use.

## EARNINGS AND ADVANCEMENT

The median annual wage for computer network architects was $120,520 in 2021. The lowest 10 percent earned less than $63,240, and the highest 10 percent earned more than $168,890.

Also in 2021, computer network architects working for insurance carriers and related activities tended to earn more, along with those working in the field of computer systems design, where salaries averaged $120,000-$127,000.

Most computer network architects work full time, and some work more than 40 hours per week. In terms of advancement, some network architects go on to become computer and information systems managers.

## EMPLOYMENT AND OUTLOOK

Employment of computer network architects is projected to grow 4 percent from 2021 to 2031, about as fast as the average for all occupations.

About 11,800 openings for computer network architects are projected each year, on average, over the decade. Many of those openings are expected to result from the need to replace workers who transfer to different occupations or exit the labor force, such as to retire.

Demand for computer network architects will grow as firms continue to design and build new information technology (IT) networks and upgrade existing ones. Continued expansion of cloud computing will increase demand for these workers to ensure that networks are properly designed and that transition to the new network proceeds smoothly.

However, demand for computer network architects is expected to be limited over the projections decade. Some computer network tasks, such as monitoring systems and troubleshooting network problems, continue to be automated and consolidated. In addition, some businesses, especially smaller ones with minimal IT requirements, find it more cost effective to contract with outside firms for these services rather than to hire computer network architects directly.

### Related Occupations

- Computer/Information Research Scientist
- Computer/Information Systems Manager
- Computer Hardware Engineer
- Computer Programmer
- Computer Support Specialist
- Computer Systems Analyst
- Database Administrator/Architect
- Information Security Analyst
- Network/Computer Systems Administrator
- Software Developer/Quality Assurance Analyst/Tester

## MORE INFORMATION

**Association for Computing Machinery (ACM)**
1601 Broadway, 10th Floor
New York, NY 10019
212.869.7440
acmhelp@acm.org
www.acm.org

**Computing Research Association (CRA)**
1828 L Street, NW, Suite 800
Washington, DC 20036
202.234.2111
info@cra.org
cra.org

**Computing Technology Industry Association (CompTIA)**
3500 Lacey Road, Suite 100
Downers Grove, IL 60515
866.835.8020
www.comptia.org

**IEEE Computer Society**
2001 L Street NW, Suite 700
Washington, DC 20036
202.371.0101
help@computer.org
www.computer.org

**National Center for Women & Information Technology (NCWIT)**
1909 26th Street, 2nd Floor
Boulder, CO 80302
303.735.6671
info@ncwit.org
www.ncwit.org

---

*Daniel Vest*

*Conversation About...*

# PRIVACY AND SOCIAL MEDIA WITH HAYLEY KAPLAN

Owner of What is Privacy?
Los Angeles, California

**How much privacy do Americans have in the online arena?**
Most people have no clue how little privacy we have. Everything we do online is tracked, unless we take steps to hide our browsing history. Our spending habits help businesses know how and what to market to us.

With a few clicks and without spending a dime, anyone can access all sorts of personal information from data broker sites. For a small fee, they can get a detailed background report with even more invasive information.

Procedures and regulations differ from state to state and some states make it remarkably easy to access official public records online. I've worked with victims of rape, assault and also victims of stalking. Online public records can make citizens vulnerable, especially if their home address is online, as in the case of an assault victim. If you make a political donation, it's online because campaigns have to be transparent. I try to educate people that every action has a consequence and we need to consciously think about that before we act.

**What are some examples of people not thinking through the consequences of online actions?**
Every Facebook profile picture is public. I know a police officer whose profile picture was a photo of her and her daughter. She thought she was doing a good job because she's monitoring her kid's account on Instagram, but her profile photo put her daughter in jeopardy.

A high school student went to a gay pride parade wearing a t-shirt with a graphic image and a crude saying on it. Years later the picture surfaced and he was ousted from his professional golfing career.

My mantra is "Anything I put online should be OK for anyone to see." People need to learn how to interact with each other without having to put it online. Just enjoy a moment. If you find yourself thinking that you must get a photo or video of everything you do to post online rather than enjoying and experiencing what you're doing, it's time to rethink things.

**It sounds like you also have to worry about what other people post.**

Absolutely. It's not hard to ruin a person's reputation nowadays. People constantly go out of their way to post bad reviews and malicious content. There are plenty of sites that welcome negative information. Also, someone can easily post a photo that you'd rather not have online, even if there is no malice intended. Sexting is very dangerous because a risqué private photo can end up being very public.

**What challenges do businesses face in the realm of privacy?**

A business that doesn't protect employee and customer data can face large fines. Businesses must take steps to prevent themselves from being hacked, and know how to recover and have the least amount of damage if they do get hacked. The rules under the General Data Protection Regulation in the EU are strict. We don't have comparable regulations—I hope we do eventually—but any company in the U.S. that markets to the EU or any international business needs to comply and know how to protect their customers' data. Marketing data that businesses collect is valuable but must be protected. It's a balancing act.

**Any other downsides to social media and the internet that students should think about?**

Cyber bullying is a horrible byproduct of the internet.

Over-posting of personal or inappropriate information can come back to haunt a person in terms of college and job acceptances.

We'll need the equivalent of AA, Alcoholics Anonymous (SMA—Social Media Anonymous?) for people addicted to social media who suffer serious repercussions as a result. Posture is awful from computers and hands are suffering from typing. Concentration and attention spans are limited as we multitask from one thing to another instead of focusing on a single activity. The ability to communicate in live situations has decreased and panic sets in when we realize we don't have our phones nearby.

One thing is for sure, the internet will continue to give rise to negative side effects, and this will open up new careers along the way for those who foresee the problems and come up with solutions.

*This interview was originally published in 2018.*

## *Conversation With...*
## RUSS WHITE

Network Architect, LinkedIn
Raleigh, North Carolina
IT professional, 30 years

**What was your individual career path in terms of education/training, entry-level job, or other significant opportunity?**

I have had many lucky breaks. That, combined with a knack for and broad background in engineering—and a lot of hard work—carried my career forward across the years. My background is actually in radio and RADAR systems, which led to discreet electronic systems, and then to computers and programming, and finally to network engineering. The entry level job that most shaped my career was in the Cisco Technical Assistance Center in Research Triangle Park in Raleigh. I joined LinkedIn last year.

**What are the most important skills and/or qualities for someone in your profession?**

It's not so much which skill; it's more a matter of learning to put every skill to some use. From writing to graphic design and tube-based radios, you can translate skills from one area to another, even if they don't seem to relate. There are three key things, however. First, you must learn how to communicate. Writing and speaking are deeply underrated skills in the network engineering world. Second, you must learn to make and keep friends. The real key to success is knowing people, no matter what you do for a living; specifically surrounding yourself with people who are much smarter than you are—not so difficult in my case most of the time! Finally, you must learn to separate principles and ideas from implementation, to understand how things work and why they work that way. You need to move beyond the engineer who knows where to hit the machine to make it work.

**What do you wish you had known going into this profession?**

I wish I'd known more about math, particularly in the realm of algorithms and processes. It's not so much being able to solve problems correctly that counts, but rather the ability to set problems up—to know what method and process you would, or could, use when facing a specific class of problem. This takes a long time to learn, and it seems almost impossible without a lot of straight experience doing it, but if there were one shortcut I wish I'd seen a lot earlier, this would be it.

**Are there many job opportunities in your profession? In what specific areas?**

I would say so, but the field is definitely running in a lot of different directions right now. The most pressing need is for "full stack" engineers, who can work on anything in or connected to the network, at least to some degree. Security is the other big deal right now.

**How do you see your profession changing in the next five years? What role will technology play in those changes, and what skills will be required?**

I hope to see network engineering move out of configuration and automation, and into trying to understand exactly how a network really works, and how it interacts with applications, traffic flows, and the like. Right now we are experiencing the pendulum swings of centralize/decentralize, vendor/not vendor, open source/open standards, and several others. What we need to do is start understanding problems and solutions in a way that allows us to actually engineer a solution, rather than just configuring or specifying our way to a solution. We need to move beyond "just make it work," to "I understand why this works."

I don't see how we can get from where we are to where we need to be, but I do think there are technology, models, and other helpers coming along that might get us over the bar.

**What do you enjoy most about your job? What do you enjoy least about your job?**

I would say what I enjoy most is finding simple ways to solve hard problems. The part I least enjoy is the constant flow of old ideas in new garb. We can only pretend an idea is new so many times. We sip from the firehose while learning very little that's actually new sometimes. This is frustrating.

**Can you suggest a valuable "try this" for students considering a career in your profession?**

There are too many to list, but perhaps the best is to try to walk through open source implementation of a routing protocol and try to understand not only how it works, but why it works that way. Or join an open standards mailing list and follow along, perhaps. There are very few ways to get into this field without actually just getting into it.

*This interview was originally published in 2016.*

# Computer Programmer

**Snapshot**

**Career Cluster(s):** Information Technology; Science, Technology, Engineering & Mathematics

**Interests:** Computers; Computer Science; Coding; Problem-solving

**Earnings (Yearly Average):** $93,000

**Employment & Outlook:** Decline Expected

## OVERVIEW

### Sphere of Work

Computer programmers write, modify, and test code and scripts that allow computer software and applications to function properly. They turn the designs created by software developers and engineers into instructions that a computer can follow. In addition, programmers run tests to ensure that newly created applications and software produce the expected results. If the products do not work correctly, programmers check the code or scripts for mistakes and modify them. This work can be exceedingly important, depending on the nature of the program being written—as in the case of cybersecurity programs.

### Work Environment

Computer programmers work in office settings, which may include working from home.

Photo via iStock.com/DragonImages. [Used under license.]

### Occupation Interest

Computer programmers are analytical and methodical by nature, and don't mind spending many hours in front of their computers, working on writing or tweaking lines of code. Moreover, they take great satisfaction in applying their knowledge of computer languages to writing programs that serve real-world functions, in many cases vital functions that factor into numerous sectors of society.

### Profile

**Working Conditions:** Mainly Inside
**Physical Strength:** Light Work
**Education Needs:** Bachelor's Degree
**Licensure/Certification:** Not Required
**Opportunities for Experience:** Internship
**Interest Score:** IC

### A Day in the Life—Duties and Responsibilities

Programmers work closely with software developers, and in some businesses their duties overlap. When such overlap occurs, programmers may be required to take on some of the tasks that are typically assigned to developers, such as designing programs.

Programmers use code libraries, which are collections of independent lines of code, to simplify their writing and improve their efficiency. They may create their own code libraries or make use of existing ones.

In addition, programmers may write or use software-as-a-service (SaaS) applications that are centrally hosted online. Although programmers typically need to rewrite their programs to work on different system platforms, such as Windows or OS X, applications created with SaaS work on all platforms. Accordingly, programmers writing SaaS applications may not have to rewrite as much code as other programmers do and can instead spend more time writing new programs.

### Duties and Responsibilities

- Writing programs in a variety of computer languages, such as C++ and Java
- Updating and expanding existing programs
- Testing programs for errors and fixing the faulty lines of computer code
- Creating, modifying, and testing code or scripts in software that simplifies development

## WORK ENVIRONMENT

### Immediate Physical Environment

Computer programmers typically work in an office setting, whether in a traditional office or at home. Regardless of the setting, they must be conscious of the health risks associated with prolonged sitting and computer-use; namely, lack of exercise and strain to eyes and hands.

Most computer programmers work full-time.

### Human Environment

Computer programmers may lead solitary work-lives, depending on their setting and the nature of the projects on which they work. More likely, however, they will need to liaise with other programmers, software developers, and engineers in the creation of a program or product. They must be able to take direction from senior programmers and project leads, and implement those ideas in practical terms.

### Technological Environment

Computer programmers must be especially tech-savvy in order to perform their duties and stay on top of evolving trends. Intimate knowledge of one or more computer languages is essential, as is the ability to work with databases, development environment software, object or component-oriented development software, web development software, and other applications such as common word processors and presentation software. Depending on the nature of the project, programmers may be required to utilize advanced and highly complex programming methods in order to create artificial intelligence (AI)-based programs, for example.

## EDUCATION & TRAINING

### High School/Secondary

High school students interested in programming as a career should take computer science-related courses, and expand their knowledge on their spare time at home. The internet is a wealth of resources about all aspects of coding and computer languages, so students should have no trouble finding this information at their fingertips, and can develop coding skills early. Interest in this subject may lead to programming as a hobby, which will serve the student well in their future career.

**Suggested High School Subjects**

- Algebra
- Applied Physics
- Biology
- Calculus
- Chemistry
- Computer Programming
- Computer Science
- Earth or Life or Physical Science

- Economics
- English
- Geometry
- History
- Political Science
- Pre-calculus
- Trigonometry

**Related Career Pathways/Majors**

*Information Technology Career Cluster*

- Programming & Software Development Pathway

*Science, Technology, Engineering & Mathematics Career Cluster*

- Engineering & Technology Pathway

### Transferable Skills and Abilities

**Analytical Skills**
- Understanding complex instructions in order to create computer code

**Communication Skills**
- Coordinating work on large projects with team members and managers

**Detail-oriented**
- Closely examining the code that they write, modify, or test, because a small mistake may affect the entire computer program

**Problem-solving Skills**
- Checking code for errors and fixing any they find

## Postsecondary

Computer programmers typically need a bachelor's degree in computer and information technology or a related field, such as mathematics. However, some employers hire workers who have other degrees or experience in specific programming languages. Programmers who work in specific fields, such as healthcare or accounting, may take classes in that field to supplement their computer-related degree. In addition, employers may prefer to hire candidates who have experience gained through internships.

Most programmers learn computer languages while in school. However, a computer science degree gives students the skills they need to learn new computer languages easily. Students get experience writing code, testing programs, fixing errors, and doing many other tasks that they will perform on the job.

**Related College Majors**

- Computer Science
- Engineering
- Information Technology
- Mathematics

## Adult Job Seekers

If an adult job seeker with no prior experience in computer programming or similar fields wishes to pursue such a career, they should enroll in a college that offers a program to expand their skills. Adults with some programming experience,

## Famous First

**In 1971, computer programming pioneer Ray Tomlinson created the first anti-virus software, Reaper. Reaper was designed to combat the Creeper program, the earliest version of a computer worm.** Source: geeksforgeeks.org

Photo by Andreu Veà (WiWiW.org), via Wikimedia Commons.

even on a hobby basis, should still find out what skills are necessary for the type of programming they wish to do on a career basis, and obtain the relevant education.

To keep up with changing technology, computer programmers may take continuing education classes and attend professional development seminars to learn new programming languages or about upgrades to programming languages they already know.

### Professional Certification and Licensure

Programmers may become certified in specific programming languages or for vendor-specific programming products. Some companies require their computer programmers to be certified in the products they use.

## EARNINGS AND ADVANCEMENT

The median annual wage for computer hardware engineers was $93,000 in 2021. The lowest 10 percent earned less than $47,560, and the highest 10 percent earned more than $155,240.

Depending on the employer, computer programmers may receive sick days, paid vacations and holidays, as well retirement benefits, and health and life insurance.

Programmers may become certified in specific programming languages or for vendor-specific programming products. Some companies require their computer programmers to be certified in the products they use.

## EMPLOYMENT AND OUTLOOK

Computer programmers held 174,400 jobs in 2021. Two percent were self-employed. Employment is projected to decline 10 percent from 2021 to 2031.

Despite declining employment, about 9,700 openings for computer programmers are projected each year, on average, over the decade. All of those openings are expected to result from the need to replace workers who transfer to other occupations or exit the labor force, such as to retire.

Computer programming can be done from anywhere in the world, so companies sometimes hire programmers in countries where wages are lower. As a result, employment growth for computer programmers in the United States may be limited. However, the high costs associated with managing projects given to overseas programmers sometimes offsets the savings from the lower wages, causing some companies to bring back or keep programming jobs in the United States.

### Related Occupations

- Computer/Information Research Scientist
- Computer/Information Systems Manager
- Computer Hardware Engineer
- Computer Network Architect
- Computer Support Specialist
- Computer Systems Analyst
- Database Administrator/Architect
- Information Security Analyst
- Network/Computer Systems Administrator
- Software Developer/Quality Assurance Analyst/Tester
- Web Developer/Digital Designer

## MORE INFORMATION

**Association for Computing Machinery (ACM)**
1601 Broadway, 10th Floor
New York, NY 10019-7440
212.869.7440
www.acm.org

**CompTIA**
3500 Lacey Road, Suite 100
Downers Grove, IL 60515
866.835.8020
membership@comptia.org
www.comptia.org

**IEEE Computer Society**
2001 L Street NW, 700
Washington, DC 20036-4928
202.371.0101
help@computer.org
www.computer.org

**National Center for Women & Information Technology (NCWIT)**
1909 26th Street, 2nd Floor
Boulder, CO 80302
303.735.6671
info@ncwit.org
www.ncwit.org

---

*Stuart Paterson; updated by Daniel Vest*

# Computer Support Specialist

## Snapshot

**Career Cluster(s):** Business Management & Administration; Information Technology; Science, Technology, Engineering & Mathematics

**Interests:** Computers & Technology; Customer Service; Data; Statistics; Investigating

**Earnings (Yearly Average):** $57,910

**Employment & Outlook:** As Fast As Average Growth Expected

## OVERVIEW

### Sphere of Work

Computer support specialists assist computer users and organizations. These specialists either maintain computer networks or provide technical help directly to computer users. Such support is especially necessary in the realm of cybersecurity, where these specialists help keep companies and their employees informed of the latest dangers and cyberattack threats, and give them essential training to combat attacks.

Photo via iStock.com/Ridofranz. [Used under license.]

### Work Environment

Some computer support specialists are able to telework. Others must be onsite or may need to travel to clients' locations.

Most computer support specialists work full time. Because computer support services may need to be available 24 hours a day, some specialists work nights or weekends.

**Profile**

**Working Conditions:** Inside
**Education Needs:** Bachelor's Degree
**Licensure/Certification:** Various
**Opportunities for Experience:** Internship
**Interest Score:** CEI

## Occupation Interest

Computer support specialists are interested in the intersection of technology and customer service. They enjoy solving problems and are good with both facts and figures, and relating to people. Methodical, thorough, and IT-savvy professionals, computer support specialists boast a mix of skills that add value to the job.

## A Day in the Life—Duties and Responsibilities

Computer network support specialists analyze and troubleshoot computer network problems. They have an important role in the daily, weekly, or monthly maintenance of their organization's networks. This maintenance may be routine or part of the organization's disaster recovery efforts. Network support specialists also may assist computer users through phone, email, or in-person visits. They often work under the direction of network and computer systems administrators, who handle more complex tasks.

**Duties and Responsibilities**

- Testing and evaluating existing network systems
- Performing regular maintenance to ensure that networks operate correctly
- Troubleshooting local area networks (LANs), wide area networks (WANs), and Internet systems
- Analyzing customers' computer problem to diagnose it and determine the cause
- Documenting customers' descriptions of their computer problems
- Guiding customers through the recommended problem-solving steps
- Setting up or repairing computer equipment and related devices
- Installing and training users on new hardware or software
- Informing team members and managers of major problems or of customers' recurring concerns

Computer user support specialists, also called help-desk technicians, usually provide technical help to non-IT computer users. They respond to requests for help in a number of ways, such as in person or by phone, online chat, or email.

Help-desk technicians solve a range of problems that vary with the industry and the particular firm. Some technicians work for large software companies or for support service firms and instruct business customers in the use of business-specific programs, such as an electronic health records program used in hospitals or physicians' offices.

Other help-desk technicians work in call centers and take customers through a problem step by step, such as to reestablish an Internet connection or to troubleshoot Wi-Fi routers or other household IT products.

## WORK ENVIRONMENT

### Immediate Physical Environment

Most computer support specialists work in offices, but some work from home. Although computer support specialists usually work in teams, some work alone. They will respond to requests for help in a number of ways, such as in person or more frequently by phone, online chat, or email.

### Human Environment

By the very nature of their job, computer support specialists work with and for others. It is a service-oriented position, so they must be comfortable interacting with people in person, on the phone, and by email.

### Technological Environment

The day-to-day technology employed by a computer support specialist tends to be routine equipment such as the telephone (or computer for email communications), connecting them to their clients seeking help. Of course, the range of knowledge possessed by these specialists is much greater than that, as they are experts in analyzing, troubleshooting and maintaining computer networks and fixing various hardware and software issues remotely.

## EDUCATION AND TRAINING

### High School/Secondary

High school students interested in becoming computer support specialists should focus on computer-related courses. Students should prepare themselves for postsecondary education, including researching the schools they would like to consider attending to achieve that goal.

**Suggested High School Subjects**

- Algebra
- Computer Programming
- Computer Science

- English
- Pre-Calculus

**Related Career Pathways/Majors**

*Business Management & Administration Career Cluster*

- General Management Pathway
- Operations Management Pathway

*Finance Career Cluster*

- Accounting Pathway
- Business Finance Pathway

*Law, Public Safety, Corrections & Security Career Cluster*

- Law Enforcement Services Pathway
- Legal Services Pathway

## Transferable Skills and Abilities

**Communication Skills**

- Clearly conveying information, both orally and in writing
- Describing solutions to computer problems in way that nontechnical users can understand

**Customer-service Skills**

- Being patient and sympathetic
- Helping people who are frustrated trying to use software or hardware

**Listening Skills**

- Understanding the problems that their customers are describing and knowing when to ask questions for clarification

**Problem-solving Skills**

- Identifying both simple and complex computer problems and then analyzing and solving them

### Postsecondary

Entry requirements vary for computer support specialists. Network support specialists typically need an associate's degree, and user support specialists typically need to complete some college courses. However, candidates for either type of position may qualify with a high school diploma plus relevant information technology (IT) certifications.

Education requirements for computer support specialists vary. Computer user support specialist jobs require some computer knowledge but not necessarily a college degree. Applicants who have taken courses in areas such as networking, server administration, and information security may qualify for these jobs. For computer network support specialists, employers may accept applicants who have an associate's degree, although some prefer that applicants have a bachelor's degree.

Large software companies that provide support to business users who buy their products or services may require applicants to have a bachelor's degree. Positions that are more technical are likely to require a degree in a field such as computer and information technology or engineering. For others, the applicant's field of degree is less important.

To keep up with changes in technology, computer support specialists may need to continue their education throughout their careers.

## Famous First

**The dot-com boom of the 1990s meant that office workers everywhere were going online, and in need of tech support. The need for superior computer support specialists is evident by the numbers: surveys show that poor customer experiences result in an estimated $83 billion loss by US enterprises each year because of defections and abandoned purchases.**

Source: thenextweb.com

Photo via iStock.com/SIphotography. [Used under license.]

**Related College Majors**

- Computer Programming
- Computer Science
- Customer Relations Management
- General Business
- Marketing

### Adult Job Seekers

Adults seeking to become computer support specialists might consider enrolling in community college or night schools where continuing education classes in computer programming are available. Although a bachelor's degree is not a requirement of this job in every instance, computer knowledge and keeping informed of the latest in technology is essential.

### Professional Certification and Licensure

Certification programs are generally offered by vendors or from vendor-neutral certification providers. Certification validates the knowledge of and best practices required by computer support specialists. Companies may require their computer support specialists to hold certifications in the products the companies use. Other types of certifications, such as CompTIA A+, may be a helpful starting point for workers seeking entry into the occupation.

## EARNINGS AND ADVANCEMENT

The median annual wage for computer network support specialists was $62,760 in 2021. The lowest 10 percent earned less than $38,560, and the highest 10 percent earned more than $102,410.

The median annual wage for computer user support specialists was $49,770 in 2021. The lowest 10 percent earned less than $34,220, and the highest 10 percent earned more than $91,060.

Most computer support specialists work full time. Because computer support services may need to be available 24 hours a day, some specialists work nights or weekends. Many computer support specialists advance to other information technology positions, such as information security analysts, network and computer systems administrators and software developers. Some become managers in the computer support services department. Some organizations provide paths for support specialists to move into other parts of the organization, such as sales.

## EMPLOYMENT AND OUTLOOK

Overall employment of computer support specialists is projected to grow 6 percent from 2021 to 2031, about as fast as the average for all occupations.

Computer network support specialists held about 185,500 jobs in 2021, while computer user support specialists held about 690,200 jobs. About 75,000 openings for computer support specialists are projected each year, on average, over the decade. Many of those openings are expected to result from the need to replace workers who transfer to different occupations or exit the labor force, such as to retire.

Computer support specialists will be needed to provide technical help and training to users with new hardware or software. However, this demand may be offset somewhat as organizations continue to implement automated tools, such as chatbots, for troubleshooting. Some businesses, especially smaller ones with minimal information technology (IT) requirements, may find it more cost effective to contract with outside firms for these services rather than to hire computer support specialists directly.

### Related Occupations

- Computer/Information Systems Manager
- Computer Network Architect
- Computer Programmer
- Computer Systems Analyst
- Customer Service Representative
- Database Administrator/Architect
- Information Security Analyst
- Network/Computer Systems Administrator
- Web Developer/Digital Designer

## MORE INFORMATION

**Association for Computing Machinery (ACM)**
1601 Broadway, 10th Floor
New York, NY 10019
212.869.7440
acmhelp@acm.org
www.acm.org

**Association of Support Professionals (ASP)**
1401 West Grand Avenue
Hot Springs, AK 71913
503.780.3520
info@asponline.com
www.asponline.com

**Computing Research Association (CRA)**
1828 L Street, NW, Suite 800
Washington, DC 20036
202.234.2111
info@cra.org
cra.org

**Computing Technology Industry Association (CompTIA)**
3500 Lacey Road, Suite 100
Downers Grove, IL 60515
866.835.8020
www.comptia.org

**Help Desk Institute (HDI)**
121 South Tejon Street, Suite 1100
Colorado Springs, CO 80903
800.248.5667
www.thinkhdi.com

**IEEE Computer Society**
2001 L Street NW, Suite 700
Washington, DC 20036
202.371.0101
help@computer.org
www.computer.org

**National Center for Women & Information Technology (NCWIT)**
1909 26th Street, 2nd Floor
Boulder, CO 80302
303.735.6671
info@ncwit.org
www.ncwit.org

**Technology & Services Industry Association (TSIA)**
17065 Camino San Bernardo, Suite 200
San Diego, CA 92127
858.674.5491
www.tsia.com

---

*Daniel Vest*

# *Conversation With...*

## JAMISON JOHNSON

Tier 2 Technical Lead, ActioNet
Washington, D.C.
Computer Support Specialist, 6 years

**What was your individual career path in terms of education, training, entry-level job, or other significant opportunity?**

I went straight into IT based on my high school curriculum and my guidance counselor's advice. I took every single computer class possible; I just liked programming classes! I received a bachelor's degree in computer information systems from Virginia State University. During my senior year, I applied, applied, applied for jobs. About three months after college, I got a consulting job with the U.S. Department of Energy as a "Tier 2" computer support specialist. Now I'm managing that team; 15 to 20 technicians report directly to me and a colleague.

We work on computer issues for the people working in our office. We handle desktop support and network connection issues. We have three tiers of help: Tier 1 Technicians are our first line of troubleshooting; if they're unable to resolve an issue in a timely manner, they escalate to us at Tier 2 to troubleshoot. Above that, Tier 3 is called in to resolve the issue and that may include network administrators, cybersecurity analysts, or website developers.

The cybersecurity threat is always there, and growing as technology advances. I hope to move into that area in the future.

**What are the most important skills and/or qualities for someone in your profession?**

In help desk roles, the key thing is your ability to troubleshoot problems, determine the cause of errors and how to fix them. You need to be an active listener and active learner. There will always be people who have been in your field longer than you; learn from them. You also have to be an instructor, because when your customers have issues, you have to teach them what to do to resolve, for example, a problem that they're causing by the way they use the computer. You need to make good, concise judgment calls.

**What do you wish you had known going into this profession?**

If I'd know that mentors are key in this profession, I would have reached out a long time ago. I try to be that person for students coming out of high school or college. Also, it's imperative to work with your colleagues and collaborate with them. Friendly competition is good, but competing all the time is not healthy.

**Are there many job opportunities in your profession? In what specific areas?**
There are many opportunities in IT, especially here in the D.C. area because of the federal government. Companies will always need people who do what I do. We resolve any customer's situation, and they will always need that support. Information security is booming, and going into this area gives you a good understanding of domestic or international cybersecurity threats. Web developers are always needed; companies will always need to market themselves online and update their websites. Once you go through one of those branches of IT, you can be a manager because somebody must manage these different groups.

**How do you see your profession changing in the next five years, what role will technology play in those changes, and what skills will be required?**
Mobile device payments, such as Apple payments, are an everyday occurrence, and we're going to see that become a norm. In order to process those payments, security needs to increase and those jobs will continue to grow. Everything from a Keurig coffee pot to your television is connected to the internet and will need to be secure. Also in the near future, your phone will have some type of universal translator device so we can bridge language barriers. Finally, we'll see a big push in robotics. If you think about it, we already have vacuum cleaners that are basically robots.

**What do you enjoy most about your job? What do you enjoy least about your job?**
I like that IT offers opportunities. Right now I'm in a help desk role, but if I work hard and network, I have options. I also like that I have early exposure to new technologies. What I least like is the lack of communication between our different operating groups. I've been with a couple of different contracting organizations and that's always an issue. We end up going through a chain of command and it circles around. A direct line of communication would be simpler.

**Can you suggest a valuable "try this" for students considering a career in your profession?**
You can easily Google any job title you want to research, or go on YouTube. A lot of companies offer internships. We've hired people with no prior experience after they interned with us. Also, you can shadow an IT professional to see if it's something you want to study long-term.

*This interview was originally published in 2016.*

# Computer Systems Analyst

**Snapshot**

**Career Cluster(s):** Arts, A/V Technology & Communications; Information Technology; Science, Technology, Engineering & Mathematics

**Interests:** Computer Science; Problem-solving; Analytical Thinking

**Earnings (Yearly Average):** $99,270

**Employment & Outlook:** Faster Than Average Growth Expected

## OVERVIEW

### Sphere of Work

Computer systems analysts, sometimes called systems architects, study an organization's current computer systems and procedures, and design solutions to help the organization operate more efficiently and effectively. They bring business and information technology (IT) together by understanding the needs and limitations of both. In terms of cybersecurity, computer systems analysts assess administrative, physical, and technical security risks for data, software, and hardware. They strive to detect, prevent, or minimize the impact of security breaches and concerns.

Photo via iStock.com/dima_sidelnikov. [Used under license.]

### Work Environment

Computer systems analysts can work directly for an organization or as contractors, often working for an information technology firm. The projects that computer systems analysts work on usually require them to collaborate and coordinate with others.

Analysts who work on contracts in the computer systems design and related services industry may move from one project to the next as they complete work for clients.

### Profile

**Working Conditions:** Inside
**Physical Strength:** Light Work
**Education Needs:** Bachelor's Degree; Master's Degree
**Licensure/Certification:** Not Required
**Opportunities for Experience:** Internship; Certification Courses
**Interest Score:** ICR

## Occupation Interest

Computer systems analysts usually have a strong interest in computer science and an aptitude for problem-solving. These professionals use logic and reasoning to assess a company's computer systems and identify areas in need of improvement. A desire to work with others is necessary in this position. Communication is an essential skill for instructing the workers within a company on the best ways to utilize their computer systems. Computer systems analysts specializing in matters of cybersecurity have additional interests in solving crimes and analyzing evidence.

### Duties and Responsibilities

- Consulting with managers to determine the role of IT systems in an organization
- Researching emerging technologies to determine if installing them can increase the organization's efficiency and effectiveness
- Preparing an analysis of costs and benefits so that management can decide if IT systems and computing infrastructure upgrades are financially worthwhile
- Devising new ways to add new functionality to existing computer systems
- Designing and implementing new systems by choosing and configuring hardware and software
- Overseeing the installation and configuration of new systems to customize them for the organization
- Conducting testing to ensure that the systems work as expected
- Training the systems' end users and writing instructions manuals

## A Day in the Life—Duties and Responsibilities

Most computer systems analysts specialize in computer systems that are specific to the organization they work with. For example, an analyst might work predominantly with financial computer systems or with engineering computer systems. Computer systems analysts help other IT team members understand how computer systems can best serve an organization by working closely with the organization's business leaders.

Computer systems analysts use a variety of techniques, such as data modeling, to design computer systems. Data modeling allows analysts to view processes and data flows. Analysts conduct in-depth tests and analyze information and trends in the data to increase a system's performance and efficiency.

Analysts calculate requirements for how much memory, storage, and computing power the computer system needs. They prepare flowcharts or other kinds of diagrams for programmers or engineers to use when building the system. Analysts also work with these people to solve problems that arise after the initial system is set up. Most analysts do some programming in the course of their work.

In some cases, analysts who supervise the initial installation or upgrade of IT systems from start to finish may be called IT project managers. They monitor a project's progress to ensure that deadlines, standards, and cost targets are met. IT project managers who also plan and direct an organization's IT department or IT policies are included in the profile of computer and information systems managers.

## OCCUPATION SPECIALTIES

**Programmer Analyst**

Programmer analysts design and update their system's software and create application tailored to their organization's needs. They do more coding and debugging than other types of analysts, although they still work extensively with management and business analysts to determine the business needs that the applications are meant to address. Other occupations that do programming are computer programmers and software developers.

**Software Quality Assurance (QA) Analyst**

Software quality assurance analysts, or QA analysts, do in-depth testing, and diagnose problems of the systems they design. Testing and diagnosis are done to make sure that critical requirements are met. QA analysts also write reports to management recommending ways to improve the systems.

## WORK ENVIRONMENT

### Immediate Physical Environment

Work settings for computer systems analysts are primarily office environments. Depending upon the company and what it specializes in, these environments may vary from job to job. For systems analysts that work individually and take on multiple clients, travel to and from multiple client locations is necessary.

### Human Environment

Computer systems analysts must have a developed skill in communication because human interaction is unavoidable in this field. To ensure a system is working properly and performing well, systems analysts must work closely with a client's IT department. If they need assistance while analyzing a computer system, they may also collaborate with other systems analysts to brainstorm a solution together.

### Technological Environment

Computer systems analysts are well-versed in computer technology and understand diverse computer software and hardware. Depending upon what is being analyzed, the software used may vary.

## EDUCATION AND TRAINING

### High School/Secondary

Computer systems analysts are usually required to have a high school diploma or equivalent. High school courses focused on science and mathematics are useful to those interested in a career in this field. Computer science courses are particularly relevant to this career path.

**Suggested High School Subjects**

- Algebra
- Biology
- Calculus
- Chemistry
- Computer Science
- Earth or Life or Physical Science
- Economics
- English
- Geometry
- History
- Physics
- Political Science
- Statistics
- Trigonometry

**Transferable Skills and Abilities**

**Analytical Skills**

- Interpreting complex information from various sources and deciding the best way to move forward on a project
- Determining how changes may affect a project

**Communication Skills**

- Working as an intermediary between management and the IT department
- Explaining complex issues in a way that both will understand

**Creativity**

- Finding innovative solutions to computer problems
- Thinking "outside the box"

**Related Career Pathways/Majors**

*Arts, A/V Technology & Communications Career Cluster*

- Telecommunications Career Pathway

*Information Technology Career Cluster*

- Information Support and Services
- Programming and Software Development

*Science, Technology, Engineering & Mathematics Career Cluster*

- Engineering & Technology Pathway

## Postsecondary

Most computer systems analysts have a bachelor's degree in a computer-related field. Because these analysts also are heavily involved in the business side of a company, it may be helpful to take business courses or major in management information systems.

Some employers prefer applicants who have a Master of Business Administration (MBA) degree with a concentration in information systems. For more technically complex jobs, a master's degree in computer science may be more appropriate.

Although many computer systems analysts have technical degrees, such a degree is not always a requirement. Many analysts have liberal arts degrees and have gained programming or technical expertise elsewhere.

**Related College Majors**

- Computer Engineering
- Computer Programming
- Computer Science
- Information Science
- Management Information Systems

## Adult Job Seekers

Computer systems analysis requires a deep knowledge of computer systems and their various applications, so interested individuals should be sure that they have the appropriate education and training. An individual with no experience in the field should enroll in a relevant program at a technical school or community college.

## Professional Certification and Licensure

Certification is not typically required by employers, though many computer systems analysts pursue additional education and certifications to remain up-to-date on the latest technologies.

**Fast Fact**

**In 2016, only three industries—government, retail, and technology—accounted for the 95% of compromised customer records. These industries remain highly sought-after targets because of the abundance of personally identifying information in their databases.** Source: TechRepublic.com

Photo via iStock.com/weerapatkiatdumrong. [Used under license.]

### Additional Requirements

Many systems analysts continue to take classes throughout their careers so that they can learn about new and innovative technologies. Technological advances come so rapidly in the computer field that continual study is necessary to remain competitive.

Systems analysts must understand the business field in which they are working. For example, a hospital may want an analyst with a thorough understanding of health plans and programs such as Medicare and Medicaid, and an analyst working for a bank may need to understand finance. If a computer systems analyst works in forensics, they should have a general understanding of law and criminal justice. Having knowledge of their industry helps systems analysts communicate with managers to determine the role of the information technology (IT) systems in an organization.

## EARNINGS AND ADVANCEMENT

Earnings depend upon the computer systems analyst's geographic location, educational attainment, and years of experience. Median annual earnings of computer systems analysts were $99,270 in 2021. The lowest 10 percent earned less than $60,680, and the highest 10 percent earned more than $158,010.

Computer systems analysts may receive benefits such as paid vacations, holidays, and sick days; life and health insurance; and retirement benefits, usually paid by the employer.

An understanding of the specific field an analyst is working in is helpful in getting a position. For example, a hospital may desire an analyst with a background or coursework in health management. Overall, candidates with a background in business may have better prospects because jobs for computer systems analysts often require knowledge of an organization's business needs.

## EMPLOYMENT AND OUTLOOK

Computer systems analysts held 538,800 jobs in 2021, and about 44,500 openings for computer systems analysts are projected each year, on average, over the decade. As organizations across the economy increase their reliance on information technology (IT), analysts will be hired to design and install new computer systems. Smaller firms with minimal IT requirements will find it more cost effective to contract with cloud service providers, or industries that employ expert IT service providers, for these workers. This contracting should lead to job growth in both the data processing, hosting, and related services industry and the computer systems design and related services industry.

Additional job growth is expected in healthcare fields. Computer systems analysts will be needed to accommodate the anticipated increase in the use and implementation of electronic health records, e-prescribing, and other forms of healthcare IT.

### Related Occupations

- Actuary
- Computer/Information Research Scientist
- Computer/Information Systems Manager
- Computer Network Architect
- Computer Programmer
- Computer Support Specialist
- Database Administrator
- Information Security Analyst
- Management Analyst
- Network/Computer Systems Administrator
- Operations Research Analyst
- Software Developer
- Web Developers and Digital Designer

## MORE INFORMATION

**Association for Computing Machinery (ACM)**
1601 Broadway, 10th Floor
New York, NY 10019-7434
212.869.7440
acmhelp@acm.org
www.acm.org

**Association of Information Technology Professionals (AITP)**
3500 Lacey Road, Suite 100
Downers Grove, IL 60515
866.835.8020
630.678.8300
www.comptia.org

**Computing Research Association (CRA)**
1828 L Street NW, Suite 800
Washington, DC 20036-4632
202.234.2111
info@cra.org
www.cra.org

**Institute for the Certification of Computing Professionals (ICCP)**
244 South Randall Road, Suite 116
Elgin, IL 60123
800.843.8227
847.299.4227
office2@iccp.org
www.iccp.org

**Institute of Electrical and Electronics Engineers Computer Society (IEEE-CS)**
2000 L Street NW, Suite 700
Washington, DC 20036-4928
202.371.0101
help@computer.org
www.computer.org

**National Center for Women & Information Technology (NCWIT)**
1909 26th Street, 2nd Floor
Boulder, CO 80302
303.735.6671
info@ncwit.org
www.ncwit.org

---

*Emily Porter; updated by Daniel Vest*

## *Conversation With...*

## MARCUS ROGERS, PHD

Professor/Assistant Dean for Cybersecurity Initiatives, Purdue University
Chief Scientist Tippecanoe County/Purdue High Tech Crime Unit
West Lafayette, Indiana
In the field, 20+ years in digital forensics

**What was your individual career path in terms of education/training, entry-level job, or other significant opportunity?**

My career path was somewhat unique. When I started, there were no academic programs in digital forensics. I started with an undergrad[uate] degree in psychology, then I became a police officer, and a detective in a computer crime unit. I continued my education and earned my master's and PhD in forensic psychology, specializing in profiling computer criminals. I was sent to various technical courses while in law enforcement. I have always been fascinated with computers and built and sold computers when I first joined the police department.

I left law enforcement and entered the consulting world where I kept taking classes and training. I joined academia in 2003 and set up one of the first formal digital forensics academic programs in the United States and continue to conduct research and teach digital forensics.

**What are the most important skills and/or qualities for someone in your profession?**

Probably the most important skills and/or qualities are:

- Attention to detail. The devil really is in the details with digital forensics. You have to be precise and accurate.
- Being curious. Always wondering what else could account for what you are seeing.
- Ability to troubleshoot. With technology and software, something will always go wrong. When it does, you need to be able to figure out how to fix it.
- Passion for learning. Technology is always changing; you need to be constantly aware of what is new and how to work with it.

**What do you wish you had known going into this profession?**

I wish I had known how slow the legal justice system is to adapt to change. Technology changes so rapidly, as does digital evidence. The courts are at least a decade behind in understanding technology and the evidence derived from computing devices. This slowness has resulted in unnecessary confusion in the justice system (both criminal and civil) as to what the limitations really are related to digital evidence.

**Are there many job opportunities in your profession? In what specific areas?**
There are far more jobs in digital forensics than there are people to fill the positions at this time (and for the foreseeable future). As we become more and more dependent on technology, digital evidence has become part of nearly every investigation. Law enforcement is overwhelmed with cases and is increasingly turning to private companies to handle the overload. Almost every civil case involves digital evidence as well.

**How do you see your profession changing in the next five years, how will technology shift, and what skills will be required?**
In the next five years, the use of the cloud and cloud computing will have a big impact on digital forensics. The current tools used in digital forensics are not well designed to deal with data that is not stored locally (i.e., on the computer or phone). Cloud computing and cloud storage will lead to an explosion in the potential volume of digital evidence and where the evidence is actually located will be uncertain, as evidence in the cloud can exist across multiple servers in multiple countries. Digital forensics will require the use of machine learning to locate and triage the vast amount of evidence so that a human investigator can make the final decision on the importance and meaning of the digital evidence.

**What do you enjoy most about your job? What do you enjoy least about your job?**
I enjoy the fact that I constantly have to learn new technologies and methods for investigating these technologies. You never get bored. I enjoy mentoring new digital forensic professionals and seeing how excited they are to be in such a dynamic forensics discipline. What I enjoy least is dealing with evidence that shows what the victim might have been going through in a case. Human suffering is hard to watch, even vicariously through digital evidence.

**Can you suggest a valuable "try this" for students considering a career in your profession?**
I strongly suggest that students job shadow someone who is [a] digital forensics professional to see what the job is really like (it is not like on the CSI TV shows). I would also suggest that they start collecting older phones and storage media (hard drives) and use some of the great open-source tools to look at the potential evidence that can be stored on these devices.

*This interview was originally published in 2021.*

## *Conversation With...*

## RADA KOGAN

Senior Programmer Analyst, Boston College
Boston, Massachusetts
IT professional, 25 years

**What was your individual career path in terms of education/training, entry-level job, or other significant opportunity?**

My path was unconventional. I never got a formal education in information technology. I got my college degree in engineering in what used to be the Soviet Union and became Ukraine. So I came here at 25 with a solid technical background because of the engineering courses and some computer science courses I had taken. It's just that "information technology" in the USSR was an oxymoron—information was controlled and censured! There was no such field as "information technology" because there was no freedom of speech. But when I came here I knew that it was something that I would like to do. I was continually educating myself, taking courses in computer science and IT at Boston University and Brandeis University, but mostly I learned on the job.

My first professional job was as a programmer analyst at Phillips Academy in Andover, MA, but I didn't do much analysis. From there, I went to the Museum of Science in Boston because they had the same system. And from there I moved to the financial sector, working for an investment company. Then I worked at Clean Harbors and later at Cognex, an engineering and manufacturing company in Natick, MA. I started at Boston College almost 10 years ago. I'm a still responsible for code development, but for the last several years, systems analysis has become a large part of my job.

**What are the most important skills and/or qualities for someone in your profession?**

You need creative thinking. You have to be independent and be detail oriented. You need the ability to interpret complex information and concepts, because you're talking to non-technical people and you have to communicate with them in their language. When they want to know why the system can't do this or can't do that, you need to be able to explain systems limitations on their level. And vice versa: you need to translate business requirements to software developers. Analysts are the go-between people. You need to be adaptable, because technology changes rapidly. You have to possess good communication skills, good presentation skills and good writing skills. Problem solving is important. You have to learn to set priorities and do project management. You need the ability to work under pressure and meet tight deadlines.

**What do you wish you had known going into this profession?**

I was at a disadvantage with limited language skills. I didn't have good writing skills. I was definitely lacking the ability to communicate to non-technical people. It was an issue because I was too technical for them. I had to learn on the job. I wish I had a better understanding of the business side. If I had a degree in information technology, it would have expanded my career choices and definitely would have helped me move up the career ladder. I strongly suggest looking at degree programs in information systems. This is what I didn't get, the business side, how to budget and calculate costs. It's extremely important to enter this field with good fundamental knowledge.

**Are there many job opportunities in your profession? In what specific areas?**

This profession will be in high demand, especially in health care, finance, and education. Every industry uses computer systems. Somebody needs to understand all these processes, how applications connect and integrate, and how information is passed between multiple systems. It's a hot field and it's going to get even hotter.

**How do you see your profession changing in the next five years? What role will technology play in those changes, and what skills will be required?**

We use PeopleSoft, which is a massive business software product. We write programs using a proprietary language; we develop applications within applications. The trend I see is decreased demand for programmers and increased demand for systems analysts. Today, software packages are highly configurable products. As a systems analyst, you need to be able to mold these products to best address the needs of your organization.

**What do you enjoy most about your job? What do you enjoy least about your job?**

I like that I do a variety of things. I interact with people. I do system design work, which I like. I do support. I do development. It's far from being mundane. I like challenges and problem solving. What I like least is the pressure and deadlines, but it's part of the job.

**Can you suggest a valuable "try this" for students considering a career in your profession?**

If you like to build diagrams and work with flow charts and you're methodical and curious and really want to understand how processes work, then systems analyst could be a good fit. You're not actually coding anything; you're doing analysis and you're doing graphs. Find an internship. This will tell you, "Oh, I actually hate doing this" or "I do like it." And find a mentor, somebody who will be willing to share their experience.

*This interview was originally published in 2016.*

# Data Scientist

## Snapshot

**Career Cluster(s):** Information Technology; Science, Technology, Engineering & Mathematics

**Interests:** Computers; Computer Science; Coding; Problem-solving

**Earnings (Yearly Average):** $100,910

**Employment & Outlook:** Much Faster Than Average Growth Expected

## OVERVIEW

### Sphere of Work

Data scientists use analytical tools and techniques to extract meaningful insights from data. These insights can then be applied and used for cybersecurity purposes.

### Work Environment

Data scientists spend much of their time in an office setting, yet a work-from-home option is often possible, as well. Most data scientists work full time.

### Occupation Interest

Data scientists have strong computing and mathematical skills and an interest in statistics, analytical work and experimenting until they find solutions to problems. They must be meticulous, highly organized and patient.

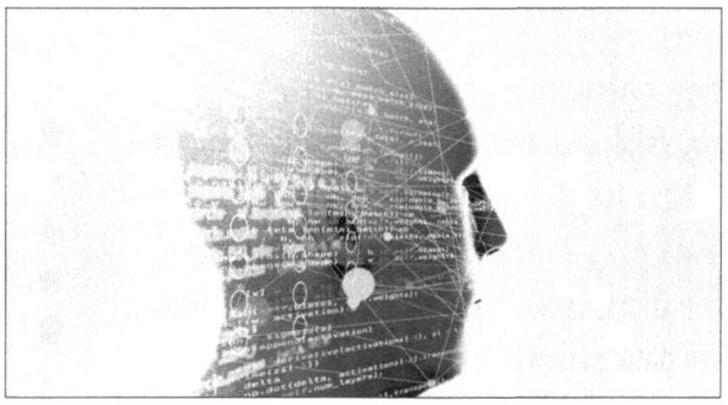

Photo via iStock.com/iambuff. [Used under license.]

## A Day in the Life—Duties and Responsibilities

### Profile

**Working Conditions:** Inside
**Education Needs:** Bachelor's Degree
**Licensure/Certification:** None
**Opportunities for Experience:** Industry-related experience or education
**Interest Score:** CEI

Data scientists often begin a project by gathering or identifying relevant data sources, such as surveys. They may use a variety of methods to obtain data, including through access to other organizations' databases or by using web-scraping tools (software that extracts specific information from websites). They may start with large, unstructured datasets, commonly referred to as raw data. To properly analyze the data, these scientists must "clean" the raw data, a process by which they structure the data to make them readable by software programs.

Data scientists develop algorithms (sets of instructions that tell computers what to do) and models to support programs for machine learning. They use machine learning to classify or categorize data or to make predictions related to the models. Scientists also must test the algorithms and models for accuracy, including for updates with newly collected data.

Data scientists often use data visualization software to present their findings as charts, maps, and other graphics. Visualization techniques allow data scientists to clearly communicate their analyses to technical and nontechnical audiences, including colleagues, managers, and clients. Ensuring that audiences understand the information helps data scientists make recommendations for business decisions or process changes based on the results of their analysis.

### Duties and Responsibilities

- Determining which data are available and useful for the project
- Collecting, categorizing, and analyzing data
- Creating, validating, testing, and updating algorithms and models
- Using data visualization software to present findings
- Making business recommendations to stakeholders based on data analysis

Some data scientists choose to focus on a particular area of work. For example, data scientists who have a strong coding or engineering background may develop or recommend systems, build machine learning algorithms, and devise ways to enhance web-browsing functions. Others conduct research for reports or academic journals. Still others focus on improving business strategy for activities such as marketing, sales, and user engagement.

## WORK ENVIRONMENT

### Immediate Physical Environment

Data scientists spend much of their time in an office setting, though some work from home.

### Human Environment

Data scientists may work in open-office environments where they exist in collaboration with other data scientists and tech professionals. They may also find themselves in instances where they have to make business recommendations to stakeholders or do other types of presentations.

### Technological Environment

In the course of their work, data scientists will employ various data-oriented programming languages and visualization software to accomplish tasks. Machine-learning skills, writing code, and being able to translate their findings for a lay audience are all part of the overall technological environment for data scientists.

## EDUCATION AND TRAINING

### High School/Secondary

High school students interested in becoming data scientists should focus on math, statistics, and computer-related courses. Becoming a data scientist oftentimes requires a bachelor's degree, so students should prepare themselves for postsecondary education, including researching the schools they would like to consider attending to achieve that goal.

**Suggested High School Subjects**

- Algebra
- Computer Science
- Computer Studies
- Computer Programming
- English
- Geometry
- Physics
- Pre-Calculus
- Statistics
- Trigonometry

**Related Career Pathways/Majors**

*Information Technology Career Cluster*

- Programming & Software Development Pathway

*Science, Technology, Engineering & Mathematics Career Cluster*

- Engineering & Technology Pathway

## Transferable Skills and Abilities

**Analytical and Critical-thinking Skills**
- Researching, examining, and interpreting findings

**Computer Skills**
- Writing code
- Analyzing data
- Developing or improving algorithms
- Using data visualization tools

**Communication Skills**
- Conveying the results of their analysis to technical and nontechnical audiences to make business recommendations.

**Logical-thinking Skills**
- Understanding and being able to design and develop statistical models and to analyze data

**Math Skills**
- Using statistical methods to collect and organize data

**Problem-solving Skills**
- Devising solutions to the problems they encounter in data collection
- Developing statistical models and algorithms

## Postsecondary

Data scientists typically need at least a bachelor's degree, but some jobs require a master's or doctoral degree. Common fields of degree include mathematics, statistics, computer science, business, and engineering.

Because data science involves the use of algorithms and statistical techniques, students need extensive study in mathematics and statistics. High school students interested in becoming data scientists should take classes in subjects such as linear algebra, calculus, and probability and statistics.

At the college level, courses in computer science are important in addition to math and statistics. Students must learn data-oriented programming languages as well as statistical, database, and other software for presenting analyses

**Related College Majors**

- Computer Science
- Engineering
- Information Technology
- Mathematics

## Adult Job Seekers

Adults seeking to become data scientists should have a degree or work background in mathematics, statistics, computer science, or a related field to enter the occupation. Continuing education may be required, so that thought should be top of mind for an adult seeking to enter the profession.

## EARNINGS AND ADVANCEMENT

The median annual wage for data scientists was $100,910 in 2021. The lowest 10 percent earned less than $59,430, and the highest 10 percent earned more than $167,040. Most data scientists work full time.

## EMPLOYMENT AND OUTLOOK

Employment of data scientists is projected to grow 36 percent from 2021 to 2031, much faster than the average for all occupations.

About 13,500 openings for data scientists are projected each year, on average, over the decade. Many of those openings are expected to result from the need to replace workers who transfer to different occupations or exit the labor force, such as to retire.

Employment growth for data scientists is expected to stem from an increased demand for data-driven decisions. The volume of data available and the potential uses for that data will increase over the projected decade. As a result, organizations will likely need more data scientists to mine and analyze the large amounts of information and data collected. Data scientists' analysis will help organizations to make informed decisions and improve their business processes, to design and develop new products, and to better market their products.

### Related Occupations

- Actuary
- Computer/Information Research Scientist

## Famous First

**Created in the 1980s, the Python programming language has become one of the most accessible programming languages due to its simple syntax, which places a greater focus on normal language, while simultaneously acting as a high-level language that can handle difficult tasks. Surveys have shown that close to 75% of data scientists use Python for their work.** Source: indatalabs.com

Image via Wikimedia Commons.

- Economist
- Financial Analyst
- Market Research Analyst
- Mathematician/Statistician
- Operations Research Analyst
- Software Developer/Quality Assurance Analyst/Tester
- Survey Researcher

## MORE INFORMATION

**Academic Data Science Alliance**
www.academicdatascience.org
Alliance for Data Science Professionals
info@afdsp.co.uk
www.alliancefordatascienceprofessionals.com

**IEEE Computer Society**
2001 L Street NW, Suite 700
Washington, DC 20036
202.371.0101
help@computer.org
www.computer.org

**Institute for Operations Research and the Management Sciences (INFORMS)**
5521 Research Park Drive, Suite 200
Catonsville, MD 21228
800.446.3676
informs@informs.org
www.informs.org

**National Center for Women & Information Technology (NCWIT)**
1909 26th Street, 2nd Floor
Boulder, CO 80302
303.735.6671
info@ncwit.org
www.ncwit.org

*Daniel Vest*

# Database Administrator/Architect

### Snapshot

**Career Cluster(s):** Information Technology

**Interests:** Computer Science; Data; Science & Technology Statistics

**Earnings (Yearly Average):** $101,000

**Employment & Outlook:** Faster Than Average Growth Expected

## OVERVIEW

### Sphere of Work

Database administrators and architects create or organize systems to store and secure a variety of data, such as financial information and customer shipping records. They also make sure that the data are available to authorized users. As so much important data is frequently stored in a database management system, database administrators and architects are tasked with protecting the core data from potential cybercriminal attacks.

### Work Environment

Database administrators and architects work in nearly all industries. For example, in retail they may design databases that track buyers' shipping information; in healthcare, they may manage databases that secure patients' medical records.

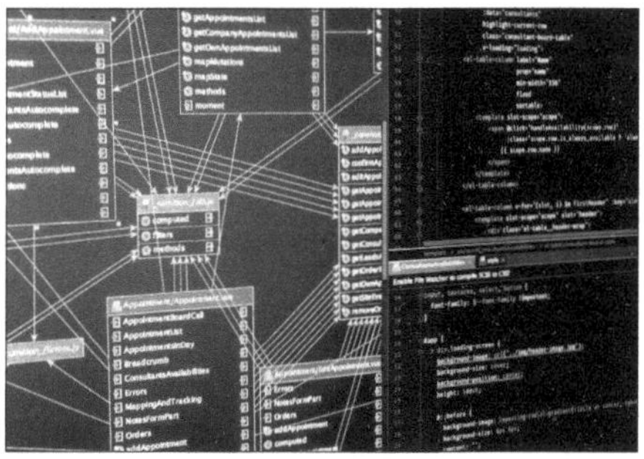

Photo via iStock.com/Yurich84. [Used under license.]

Most database administrators and architects work full time.

## Occupation Interest

Database architects and administrators value independent work that is technical in nature, using their knowledge to better serve user needs through the organization and implementation of systems.

### Profile

**Working Conditions:** Inside
**Education Needs:** Bachelor's Degree
**Licensure/Certification:** Vendor-provided; varies by employer and software type
**Opportunities for Experience:** Internship
**Interest Score:** ICE

## A Day in the Life—Duties and Responsibilities

Database administrators, often called DBAs, make sure that data analysts and other users can easily use databases to find the information they need. They also ensure that systems perform as they should by monitoring database operation and providing support.

Many databases contain personal, proprietary, or financial information. Database administrators often are responsible for planning security measures to protect this important information.

Database architects design and build new databases for systems and applications. They research the technical requirements of an organization during the design phase and then create models for building the database. Finally, they code new data architecture, integrating existing databases or infrastructure, and check for errors or inefficiencies.

### Duties and Responsibilities

- Identifying user needs
- Creating and administering databases
- Ensuring that organizational data are secure
- Backing up and restoring data to prevent data loss
- Ensuring that databases operate efficiently and without error
- Making and testing modifications to database structure when needed
- Maintaining databases and updating permissions

The duties of database administrators and database architects may overlap. For example, administrators and architects may be generalists who work on both systems and applications. However, some DBAs specialize in certain tasks, such as maintenance, that vary with an organization and its needs.

## OCCUPATION SPECIALTIES

### Application DBA

Application DBAs do all the tasks of a general DBA focusing solely on a database for a specific application or set of applications, such as customer-service software. They may write or debug programs and must be able to manage the applications that work with the database.

**System DBA**

System DBAs are responsible for the physical and technical aspects of a database, such as installing upgrades and patches to fix program bugs. They ensure that the firm's database management systems work properly.

## WORK ENVIRONMENT

### Immediate Physical Environment

Most database architects and administrators work in office environments, be it in a traditional setting amongst other professionals, or in a home-office setting, which can sometimes involve virtual interactions.

### Human Environment

In this job, database administrators and architects will come into contact with other developers, other workers and higher-ups such as management, and will need to possess a degree of people skills, and communicate their ideas in a clear and precise way, in a professional manner.

### Technological Environment

A high degree of technological-savvy is required, as database administrators and architects must take on jobs with high degrees importance to the organizations they work for, such as coding new data architecture, integrating existing databases or infrastructure, and check for errors or inefficiencies.

## EDUCATION AND TRAINING

### High School/Secondary

High school students interested in becoming database administrators or architects should focus on math, statistics, and computer-related courses. An entry into this profession requires at least a bachelor's degree, so students should prepare themselves for postsecondary education, including researching the schools they would like to consider attending to achieve that goal.

**Suggested High School Subjects**

- Algebra
- Computer Science
- Chemistry
- Geometry
- Physics
- Pre-Calculus
- Statistics
- Trigonometry

**Related Career Pathways/Majors**

*Information Technology Career Cluster*

- Network Systems Pathway
- Information Support & Services Pathway

*Science, Technology, Engineering & Mathematics Career Cluster*

- Engineering & Technology Pathway
- Science & Mathematics Pathway

### Transferable Skills and Abilities

**Analytical Skills**

- Monitoring a database system's performance to determine when action is needed
- Evaluating information from a variety of sources
- Deciding on an approach

**Communication Skills**

- Conveying information effectively to developers, managers, and other workers

**Detail-oriented**

- Possessing an understanding of complex systems, in which a minor error can cause major problems

**Problem-solving Skills**

- Troubleshooting and correcting problems as they arise

## Postsecondary

Database administrators and architects typically need a bachelor's degree in computer and information technology or a related field, such as engineering; some DBAs study business. Employers may prefer to hire applicants who have a master's degree focusing on data or database management, typically either in computer science, information systems, or information technology.

Database administrators and architects need an understanding of database languages, such as Structured Query Language, or SQL. DBAs will need to become familiar with whichever programming language their firm uses.

**Related College Majors**

- Applied Mathematics
- Applied Statistics
- Computer Engineering
- Computer Science
- Data Science

### Adult Job Seekers

A bachelor's degree in computer and information technology or a related field is usually a necessity for anyone wanting to enter this profession, be they adult job seekers or younger. Some DBAs study business prior to entering the field.

### Professional Certification and Licensure

Certification is typically offered directly from software vendors or vendor-neutral certification providers. Employers may require their database administrators and architects to be certified in the products they use.

## EARNINGS AND ADVANCEMENT

The median annual wage for database administrators was $96,710 in 2021. The lowest 10 percent earned less than $48,880, and the highest 10 percent earned more than $151,400.

The median annual wage for database architects was $123,430 in 2021. The lowest 10 percent earned less than $63,260, and the highest 10 percent earned more than $169,500.

Database administrators and architects may advance to become computer and information systems managers. Experienced database administrators may advance to become database architects.

## EMPLOYMENT AND OUTLOOK

Overall employment of database administrators and architects is projected to grow 9 percent from 2021 to 2031, faster than the average for all occupations.

About 11,500 openings for database administrators and architects are projected each year, on average, over the decade. Many of those openings are expected to result from the need to replace workers who transfer to different occupations or exit the labor force, such as to retire.

Employment growth will be driven by the continued data needs of companies in nearly all sectors of the economy. Database administrators and database architects will be needed to organize and present information to stakeholders in a user-friendly format. As organizations continue to migrate to cloud environ-

## Fast Fact

**"Cyber ranges" are virtual environments used for training in cybersecurity, cyberwarfare, simulating or emulating cybersecurity scenarios, and developing cybersecurity-related technology. Their size can be quite variable, ranging from a single node to a network resembling the internet.** Source: nist.gov

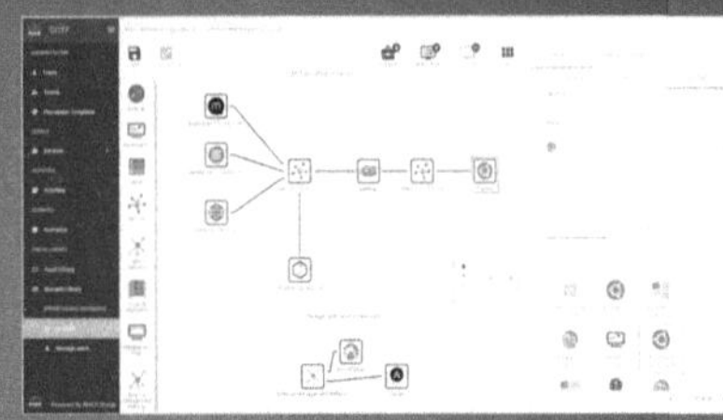

Image by Covelus, via Wikimedia Commons.

ments, these administrators and architects will be critical to ensuring proper database design, transition, backup, and security and to ensuring that connections to legacy systems remain intact.

### Related Occupations

- Computer/Information Systems Manager
- Computer Network Architect
- Computer Programmer
- Computer Support Specialist
- Computer Systems Analyst
- Financial Analyst
- Information Security Analyst
- Market Research Analyst
- Network/Computer Systems Administrator
- Operations Research Analyst
- Software Developer/Quality Assurance Analyst/Tester
- Web Developer/Digital Designer

## MORE INFORMATION

**Academic Data Science Alliance**
www.academicdatascience.org
Alliance for Data Science Professionals
info@afdsp.co.uk
www.alliancefordatascienceprofessionals.com

**Association for Computing Machinery (ACM)**
1601 Broadway, 10th Floor
New York, NY 10019
212.869.7440
acmhelp@acm.org
www.acm.org

**Computing Research Association (CRA)**
1828 L Street NW, Suite 800
Washington, DC 20036
202.234.2111
info@cra.org
www.cra.org

**IEEE Computer Society**
2001 L Street NW, Suite 700
Washington, DC 20036
202.371.0101
help@computer.org
www.computer.org

**Institute for Operations Research and the Management Sciences (INFORMS)**
5521 Research Park Drive, Suite 200
Catonsville, MD 21228
800.446.3676
informs@informs.org
www.informs.org

**National Center for Women & Information Technology (NCWIT)**
1909 26th Street, 2nd Floor
Boulder, CO 80302
303.735.6671
info@ncwit.org
www.ncwit.org

---

*Daniel Vest*

## *Conversation With...*
## LEONORA GIOVANANGELO

Executive Office of Labor & Workforce Development
Commonwealth of Massachusetts
IT professional, 29 years
Database Management, 16 years

**What was your individual career path in terms of education/training, entry-level job, or other significant opportunity?**

I was studying accounting at Bentley College in Massachusetts and I was not passionate about it. We had to take an IT class as a requirement. I just fell in love with programming because I'm a very black-and-white person. I love math because to me, there's one right answer. I have a very mathematical mind. It's very easy for me to figure out if I've done something well. I have a passion for IT. After college, I started out in my first job as a programmer, which I did for three or four years. Then an opportunity come up to be a database administrator (DBA) and I jumped on it.

A database is a collection of information; it's basically where the information is held. It's almost like a big Excel spreadsheet. A lot of being a database administrator is just making sure that everything is put together as efficiently as possible, so that the information can be accessed efficiently. It's all about relationships between the data. I've been working for the state of Massachusetts since 2002. Before that, I was a database administrator for Boston University. Working for the state and working for private enterprise are very different. The pace is very different. Things get done more quickly in the private sector. In the private sector one can expect to put in extra hours, whereas in the public sector, due to union rules, your days typically have a specific start and end time.

**What are the most important skills and/or qualities for someone in your profession?**

Attention to detail is big, and the ability to solve problems. You also need to be able to think outside the box. Often when programmers and developers have a problem, their first line of defense will be to talk to the DBA. I see myself as being a business analyst. Data warehousing is for reporting information. You have to meet with users and understand their terminology and really, really understand what they do and then report back to the developers. A DBA is like a middleman between business and IT. Being able to speak both sides is critical. If you don't understand the business side, you're not going to get the right answers. You need to be able to ask the right questions—and that goes for IT in general. You need to be able to think quickly. Communication skills and the ability to get along with other people is key, in any job. If you don't have that, you're in trouble.

**What do you wish you had known going into this profession?**
How demanding it can be. You're on call on the time. I've been on call 24 hours a day, 7 days a week, 365 days a year. I missed out on a lot of things when I was over at BU. You have to be flexible, that's for sure. Some DBA jobs don't require that, because they don't have off-hours support. Sometimes when you're on call, you can handle it without going into the office. It depends on how serious it is. A representative from each group might have to come in and be onsite. When I was younger, I thought that was exciting. Also, when you're younger, you have a lot f energy. But now my priorities have changed. It's about quality of life.

**Are there many job opportunities in your profession? In what specific areas?**
Yes, there are. Database administration is always gong to be hot. Every company needs a database administrator—at least one DBA. We have several here. Other big areas are security, software engineering, software developer, project management and web developer.

**How do you see your profession changing in the next five years? What role will technology play in those changes, and what skills will be required?**
Security is huge. It's a server environment here and every week the help desk sends out emails about some sort of security concern. Working in IT, you have to keep a pulse on what's out there and keep yourself current, or you can really be left behind.

**What do you enjoy most about your job? What do you enjoy least about your job?**
What I enjoy most is solving problems. I'm a real problem solver. I really enjoy getting out from behind the desk staring at my computer and talking to people. There are people in IT who thrive on that, because they can't deal with people. But as time goes on, I'm realizing that I really enjoy interacting with other people.

What I like least is being on call.

**Can you suggest a valuable "try this" for students considering a career in your profession?**
A fantastic resource is YouTube. There's a lot of information there to help you figure out whether or not you like something. On coursera.org you can take free courses from top universities. Microsoft Virtual Academy (mva.microsoft.com) offers free online training, including a section for students. If working on a spreadsheet is something that catches your attention, then, yes, database management might be a good option for you.

*This interview was originally published in 2016.*

# Financial Analyst

**Snapshot**

**Career Cluster(s):** Business Management & Administration; Finance

**Interests:** Accounting; Finance; Data; Statistics; Recordkeeping; Investigating

**Earnings (Yearly Average):** $95,570

**Employment & Outlook:** Faster Than Average Growth Expected

## OVERVIEW

### Sphere of Work

Financial analysts guide businesses and individuals in decisions about expending money to attain profit. They assess the performance of stocks, bonds, and other types of investments. Due to the sensitive nature of the information they analyze and recommend, cybersecurity is an important factor.

### Work Environment

Financial analysts work in offices, but may travel for work to various companies and clients.

### Occupation Interest

Financial analysts are analytically minded, good decision-makers under pressure, and have a fascination with crunching numbers. Proficient with technology and great communicators, they are able to combine a love of facts and figures with a love of helping others.

Photo via iStock.com/Wavebreakmedia. [Used under license.]

### A Day in the Life—Duties and Responsibilities

Financial analysts evaluate opportunities to commit money for the purpose of generating profit.

Financial analysts can be divided into two categories: buy-side analysts and sell-side analysts.

## Profile

**Working Conditions:** Inside
**Education Needs:** Bachelor's Degree; Master's Degree
**Licensure/Certification:** Various; FINRA and CFA
**Opportunities for Experience:** Business Degree; Master's Degree
**Interest Score:** CEI

Buy-side analysts develop investment strategies for companies that have a lot of money to invest. These companies, called institutional investors, include hedge funds, insurance companies, independent money managers, nonprofit organizations with large endowments, private equity firms, and pension funds.

Sell-side analysts advise financial services sales agents who sell stocks, bonds, and other investments.

Analysts may work for the business media or other research houses, which are independent from the buy and sell side.

Financial analysts generally focus on trends affecting a specific geographical region, industry, or type of product. For example, they may focus on a subject area or a foreign exchange market. They must understand how economic trends, new regulations, policies, and political situations may affect investments.

Investing has become more global, and some specialize in a particular country or world region. Companies want these specialists to understand the business environment, culture, language, and political conditions in the country or region that they cover.

## Duties and Responsibilities

- Recommending individual investments and collections of investments, known as portfolios
- Evaluating current and historical financial data
- Studying economic and business trends
- Examining a company's financial statements to determine its value
- Meeting with company officials to gain better insight into the company's prospects
- Assessing the strength of the management team
- Preparing written reports

## OCCUPATION SPECIALTIES

### Financial Risk Specialist

Financial risk specialists, also called financial risk analysts, evaluate threats to investment decisions and determine how to manage unpredictability and limit potential losses. They make investment decisions such as selecting dissimilar stocks or having a combination of stocks, bonds, and mutual funds in a portfolio. They also make recommendations to limit risk.

### Fund Manager

Fund managers work exclusively with hedge funds or mutual funds. Both fund managers and portfolio managers frequently make buy or sell decisions in reaction to quickly changing market conditions.

### Investment Analyst

Investment analysts assess information involving investment programs or financial data of institutions, such as business valuation. They also respond to queries from clients and client advisors regarding asset allocation and alternative investment topics including hedge funds, real property, and venture capital.

### Portfolio Manager

Portfolio managers select the mix of products, industries, and regions for their company's investment portfolio. These managers are responsible for the overall performance of the portfolio. They are also expected to explain investment decisions and strategies in meetings with stakeholders.

### Ratings Analyst

Ratings analysts evaluate the ability of companies or governments to pay their debts, including bonds. Based on these evaluations, a management team rates the risk of a company or government not being able to repay its bonds.

### Securities Analyst

Securities analysts evaluate securities markets and trends to identify high-yield assets for clients and companies. They may use resources such as bond performance reports, daily stock quotes, market and economic forecasts, and other financial statements and publications.

## WORK ENVIRONMENT

### Immediate Physical Environment

Most financial analysts work in offices. Some travel may be required, as they meet with company officials to offer an analysis of data, take meetings and deliver reports.

### Human Environment

Financial analysts may travel to visit companies or clients, and therefore it is not just a "numbers" career: strong oral and written communication skills are essential to the job at hand.

### Technological Environment

Financial analysts may use technology to increase their productivity. Automating some routine tasks makes these workers more efficient by allowing them to focus on analysis and other high-level responsibilities.

## EDUCATION AND TRAINING

### High School/Secondary

High school students interested in becoming financial analysts should focus on math, statistics, and business-related courses. This job requires at least a bachelor's degree, so students should prepare themselves for postsecondary education, including researching the schools they would like to consider attending to achieve that goal.

**Suggested High School Subjects**

- Algebra
- Economics
- Geometry
- Pre-Calculus
- Statistics

**Related Career Pathways/Majors**

*Business Management & Administration Career Cluster*

- General Management Pathway
- Operations Management Pathway

*Finance Career Cluster*
- Accounting Pathway
- Business Finance Pathway
- Securities & Investments Pathway

## Postsecondary

Most entry-level positions for financial analysts require a bachelor's degree; a common field of degree is business. Some employers prefer to hire job candidates who have a master's degree.

### Related College Majors
- Accounting
- Auditing
- Business Management & Administration
- Economics
- Financial Management
- General Business
- Insurance
- International Business
- Mathematics

### Transferable Skills and Abilities

**Analytical and Critical-thinking Skills**
- Evaluating a range of information in finding profitable investments

**Communication Skills**
- Clearly explaining their recommendations to clients

**Computer Skills**
- Being adept at using software
- Analyzing financial data and trends
- Creating portfolios
- Making forecasts

**Decision-making Skills**
- Reaching conclusions so that they can recommend whether to buy, hold, or sell a security.

**Detail-oriented**
- Paying attention when reviewing a possible investment, as even small issues may have large implications for its health

**Math Skills**
- Using mathematics to estimate the value of financial securities

## Adult Job Seekers

Adults seeking to become financial analysts should have a background or strength in math or finance. A bachelor's degree in one of these related disciplines is usually required, so thought should be given to an individual's long-term professional goals.

## Professional Certification and Licensure

The Financial Industry Regulatory Authority (FINRA) is the main licensing organization for the securities industry. A license is generally required to sell financial products, which may apply to some positions. Because most of the licenses require sponsorship by an employer, companies do not expect individuals to have these licenses before starting a job.

Employers often recommend certification, which may improve the chances for advancement. An example is the Chartered Financial Analyst (CFA) certification

## Fast Fact

**According to the 2019 "Cost of Cybercrime Study in Financial Services" report by Accenture, the average number of breaches in the financial sector jumped by 13%, from 134 in 2017 to 152 in 2018. Overall, the average annual cost of cybercrime per company within the financial sector was USD 18.5 million.** Source: stealthlabs.com

Photo via iStock.com/Rawpixel. [Used under license.]

from the CFA Institute. Financial analysts can become CFA certified if they have a bachelor's degree and several years of work experience and pass multiple exams. They also may choose to become certified in their field of specialty.

## EARNINGS AND ADVANCEMENT

The median annual wage for financial and investment analysts was $91,580 in 2021. The lowest 10 percent earned less than $57,900, and the highest 10 percent earned more than $166,560.

The median annual wage for financial risk specialists was $100,000 in 2021. The lowest 10 percent earned less than $59,370, and the highest 10 percent earned more than $171,000.

Fund managers are typically compensated by fees, usually structured as a percentage of assets under management and a percentage of the fund's annual return.

Financial analysts typically start by specializing in an investment field. As they gain experience, they may become portfolio managers and select the mix of investments for a company's portfolio. They also may become fund managers of large investment portfolios for individual investors. Having a master's degree in finance or business administration may improve an analyst's chances of advancing to one of these positions.

## EMPLOYMENT AND OUTLOOK

Overall employment of financial analysts is projected to grow 9 percent from 2021 to 2031, faster than the average for all occupations.

About 32,000 openings for financial analysts are projected each year, on average, over the decade. Many of those openings are expected to result from the need to replace workers who transfer to different occupations or exit the labor force, such as to retire.

Demand for financial analysts generally increases with overall economic activity. These workers will be needed to evaluate investment opportunities when new businesses are established or as existing businesses expand. In addition, emerging markets throughout the world are providing new investment opportunities, requiring expertise in geographic regions where those markets are located.

Demand also is projected to increase as big data and technological improvements allow financial analysts to conduct high-quality analysis. This analysis will help businesses manage their finances, identify investment trends, and deliver new products or services to clients.

### Related Occupations

- Budget Analyst
- Financial Manager
- Insurance Manager
- Personal Financial Advisor
- Securities/Commodities/Financial Services Sales Agent

## MORE INFORMATION

**Association for Financial Professionals (AFP)**
4520 East-West Highway, Suite 800
Bethesda, MD 20814
301.907.2862
customerservice@AFPonline.org
www.afponline.org

**CFA Institute**
1401 New York Avenue NW
Washington, DC 20005
434.951.5499
info@cfainstitute.org
www.cfainstitute.org

**Financial Industry Regulatory Authority (FINRA)**
1735 K Street NW
Washington, DC 20006
301.590.6500
www.finra.org

**Global Academy of Finance and Management (GAFM)**
P.O. Box 293
1670F East Cheyenne Mountain Boulevard
Colorado Springs, CO 80906
www.gafm.com

---

*Daniel Vest*

## *Conversation With...*
# CLIFFORD ROSSI, PHD

Professor of Finance & Executive in Residence
University of Maryland, College Park, Maryland
Financial risk manager, 25 years
Professor of finance, 7 years

**What was your individual career path in terms of education/training, entry-level job, or other significant opportunity?**

I double majored in economics and public policy at the University of Denver thinking I would have a career in government. I graduated during a recession and went to work for a small solar energy research center as a purchasing manager. Two people there steered me to Cornell University, where I earned a master's degree and a PhD in quantitative finance.

I never needed a PhD to do the job I ended up doing: financial risk management. But my doctorate gave me the structure to think critically in a deep way. Several years into my career, management realized I could translate the technical to a non-technical audience. That made me a valuable commodity. After that, my career went a completely different way; I went into management and rose up the ranks. I advise my students: find what differentiates you from the competition.

After earning my PhD, I started in government at the U.S. Treasury Department, moved into quasi-government doing risk management at mortgage-finance programs Fannie Mae and Freddie Mac, then went into private banking. A pivotal moment came in 1996 when Freddie Mac asked me to write an automated underwriting scorecard. I convinced the Federal Housing Administration commissioner that this was a great way to go. That's when management realized: This guy shouldn't just do statistical work; he needs to be a voice for how to use these models.

By the end of 2003, I had plateaued. A headhunter helped me get into the private sector, working for Countrywide Mortgage. It was a relatively smaller institution at the time and they were looking to grow significantly. I had to create their risk management function. When I left, I was Managing Director and Chief Risk Officer.

I moved on to Washington Mutual (WaMu)—then the country's largest savings and loan—as Chief Credit Officer, but was there only seven months. I left because I was dissatisfied with the bank's direction. This was 2007, as the financial crisis unfolded. I had interviewed with Citigroup when I interviewed at WaMu, and they still wanted me to work for them. It was a good move. Soon after, WaMu suffered the largest failure in the nation's financial history.

I became Managing Director and Chief Risk Officer for the Consumer Lending Group. It was the best job I ever had, overseeing 700 people at a prestigious institution. I came on under a new CEO who embraced risk management as he guided the bank through the crisis and its aftermath. I stayed two years, working out of the headquarters in New York. This was a highly stressful time. My job also became more global and demanded extensive travel.

My banking career gave my wife and me the earnings and flexibility to decide how we wanted to live our lives, so when an interesting opportunity arose to shift gears and return home, I took it.

In 2009, I joined the faculty of the University of Maryland's Robert H. Smith School of Business as a full-time professor. Now, I teach graduate students part-time and have become Chief Economist for Radian Guarantee, a large mortgage insurance company in Philadelphia.

**What are the most important skills and/or qualities for someone in your profession?**
You need some degree of analytical skill, critical thinking skills, and financial software skills—the ability to mine or extract data and manipulate and analyze it. You have to be able to communicate effectively and you need to be a team player. Risk management is an art and a science, though, so you've also got to have intuition, knowledge and judgment.

**What do you wish you had known going into this profession?**
I wish I had known that risk management wasn't viewed as an equal partner in the banking profession at the time I was there. It's become more of a focus for banks since the crisis.

**Are there many job opportunities in your profession? In what specific areas?**
If any area of banking is growing, it's risk management. It's a great career. It's at the vortex of everything a bank is about: strategy, analytics, consumers, wealth management—you can work in any part of the business.

**How do you see your profession changing in the next five years, what role will technology play in those changes, and what skills will be required?**
With the advent of high-end computing software, the risk management function has become highly quantitative and model-based and the profession is more aligned with actuarial science. Technology will continue to make our profession more quantitative.

**What do you enjoy most about your job? What do you enjoy least about your job?**
In my career in private finance, I enjoyed improving our bank's risk profile and infrastructure.

I detested some of the power-play issues that surround high-level management, such as passive-aggressive behavior from peers, or being ignored. I don't suffer fools lightly.

**Can you suggest a valuable "try this" for students considering a career in your profession?**
Attend an event at the local chapter of a professional association for risk management. People are likely to share their stories and help you understand what it is to be a financial risk manager.

*This interview was originally published in 2016*

# Financial Examiner

**Snapshot**

**Career Cluster(s):** Finance; Government & Public Administration

**Interests:** Financial Planning; Analyzing Data; Accounting; Decision-making; Mathematics

**Earnings (Yearly Average):** $81,410

**Employment & Outlook:** Much Faster Than Average Growth Expected

## OVERVIEW

### Sphere of Work

Financial examiners ensure compliance with laws that govern institutions managing monetary transactions. They review balance sheets, evaluate the risk level of loans, and assess bank management. Financial examiners typically work in one of two main areas: risk assessment or consumer compliance.

In terms of cybersecurity, financial examiners must also use their investigative and analytical skills to determine if there is compliance within their organization related to the storage and protection of customer and institutional data. Their expertise is valuable in assessing an organization's susceptibility to fraud in addition to providing direction to government agencies in criminal investigations.

Photo via iStock.com/RossHelen. [Used under license.]

### Work Environment

Financial examiners typically work in offices and spend much of their time using computers and other digital tools to evaluate data and prepare presentations and financial reports for their clients or employers. They often must travel to inspect a bank on-site. Depending upon the nature of their assignment, they will find themselves working both independently and directly with others; these professionals must be comfortable with both work environments. Financial examiners typically work regular, full-time hours.

### *Profile*

**Working Conditions:** Inside
**Physical Strength:** Light Work
**Education Needs:** Bachelor's Degree
**Licensure/Certification:** Not Required
**Opportunities for Experience:** Internship; Certification Courses; On-the-Job Training
**Interest Score:** EC

## Occupation Interest

Financial examination can be considered a branch of accounting that deals with federal and state regulations and financial institutions. Whether working in risk scoping or consumer compliance, the obligation of financial examiners is to prevent financial institutions from engaging in activities that can negatively affect clients or investors, therefore, financial examiners should have an interest in the public's welfare. Individuals in this occupation are also typically interested in finance, mathematics, and research.

## A Day in the Life—Duties and Responsibilities

Financial examiners usually work in one of two main specializations: risk assessment or consumer compliance.

### *Duties and Responsibilities*

- Monitoring the conditions of banks and other financial institutions
- Reviewing balance sheets, operating income and expense accounts, and loan documentation to confirm an institution's assets and liabilities
- Preparing reports that detail an institution's safety and soundness
- Examining the minutes of meetings of managers and directors
- Training other examiners in the financial examination process
- Reviewing and analyzing new regulations and policies to determine their impact on an institution
- Establishing guidelines for procedures and policies that comply with new and revised regulations

Those working in risk assessment evaluate the health of financial institutions. Their role is to ensure that banks and other financial institutions offer safe loans and that they have enough cash on hand to manage unexpected losses. These procedures help ensure that the financial system remains stable. These examiners also evaluate the performance of bank managers.

Financial examiners working in consumer compliance monitor lending activity to ensure that borrowers are treated fairly. They ensure that banks extend loans that borrowers are likely to be able to pay back. They help borrowers avoid "predatory loans"—loans that may generate profit for banks through high interest payments but may be costly to borrowers and damage their credit scores. Examiners also ensure that banks do not discriminate against borrowers based on race, ethnicity, or other characteristics.

Financial examiners functioning in the role of cybersecurity examiners must determine the effectiveness of an institution's cybersecurity risk identification and management processes, ensuring compliance with applicable banking laws, regulations, and policy statements.

## WORK ENVIRONMENT

### Immediate Physical Environment

Financial examiners typically do most of their work in office environments utilizing technological tools. In some cases, examiners may travel to different financial institutions to analyze data on-site or to interview members of an organization's staff, while in other cases examiners may remain in a single office and review data sent to them from institutions undergoing an examination.

### Human Environment

Financial examiners can work alone or with other examiners and financial specialists depending upon the situation. They may be required to meet with clients to obtain information for an analysis of a company's finances and policies. While conducting the analysis, the work may be more isolated. However, while leading the discussion on their findings, direct human interaction is obligatory. Interaction and cooperation with law enforcement is expected in the field of forensics and criminal justice.

### Technological Environment

Financial examiners utilize an array of computer programs and software to conduct financial analyses. Examiners should be proficient in various accounting software or able to understand and learn how to utilize these financial analysis programs quickly. Specialized software programs, such as Oversight Insights On Demand or the System for Electronic Rate and Form Filing (SERFF), are made specifically for monitoring compliance to the law and ethical communication with the consumer in a financial institution. Spreadsheet, word processing, project management, presentation, and other office software are important in this profession for creating and delivering reports to clients, or in the case of forensics, to litigation teams and investigative law enforcement.

## EDUCATION AND TRAINING

### High School/Secondary

For high school students interested in a career as a financial examiner, they should participate in courses that help prepare for a bachelor's degree in finance, accounting, or a related field. Classes in mathematics, business, economics, and accounting—if available—are the most beneficial to this profession. In addition, subjects that improve written and verbal communication skills such as public speaking, English, and writing are favorable.

**Suggested High School Subjects**

- Algebra
- Biology
- Calculus
- Chemistry
- Computer Science
- Earth or Life or Physical Science
- Economics
- English
- Environmental Science
- Foreign Language
- Geometry
- History
- Physics
- Psychology
- Public Speaking
- Sociology
- Statistics

***Transferable Skills and Abilities***

**Analytical Skills**

- Evaluating how well the managers of financial institutions are managing risk and whether the individual loans the institution makes are safe

**Detail-oriented**

- Paying close attention to minutiae when reviewing balance sheets in order to identify risky assets

**Math Skills**

- Calculating and monitoring balance sheets to ensure that a financial institution has available cash

**Writing Skills**

- Writing reports on the safety and soundness of financial institutions
- Explaining technical information clearly

**Related Career Pathways/Majors**

*Finance Career Cluster*

- Business Finance Pathway

*Government & Public Administration Career Cluster*

- Regulation Pathway

### Postsecondary

Financial examiners typically need a bachelor's degree. Although a specific major is usually not required, examiners generally need some coursework in accounting, business, finance, or a related field. Those looking to advance in the field or to become financial examination supervisors or administrators, might also consider pursuing a postgraduate degree in finance or accounting. Classes in economics, governance, and politics will also be helpful as examiners are charged with balancing institutional goals and public policy. Students interested in being a forensic financial examiner can expand their skill set by taking courses in criminal justice and law.

**Related College Majors**

- Accounting
- Business
- Economics
- Finance

## Adult Job Seekers

Adult job seekers interested in financial examination that currently hold a degree in a field related to finance or accounting can apply to open positions. On-the-job training is typically provided in these positions, so previous experience is not a direct requirement. However, individuals without a degree related to this field or experience in a similar occupation should consider returning to a continuing education program. Familiarity with accounting is a necessity, so taking additional courses in this subject and similar subjects will aid in pushing an adult's finance career forward.

## Professional Certification and Licensure

Although it is not required, professional certification indicates competencies for financial examiners who have it. The Society of Financial Examiners (SOFE) offers the Accredited Financial Examiner (AFE) and the Certified Financial Examiner (CFE) designations. Both may be earned after completing extensive requirements and passing a series of examinations. Continuing education is required to maintain these designations.

Some financial examiners become Certified Public Accountants (CPAs). CPAs are licensed by their state's Board of Accountancy. Becoming a CPA requires passing a national exam and meeting other state requirements.

## Additional Requirements

Financial examiners are required to have excellent skills in mathematics and analysis that are needed for examining an organization's financial practices and history. Proficiency in the computer programs used daily is not required because these can be learned through on-the-job training and hands-on experience. However, a proficiency in communication is imperative in this profession to ensure that professionals can report their findings clearly.

## EARNINGS AND ADVANCEMENT

Median annual earnings of financial examiners were $81,410 in 2021. The lowest 10 percent of examiners earned less than $48,830, and the highest 10 percent earned more than $160,850.

After obtaining work experience, an individual can advance in the field by applying for higher-paying positions in other organizations or by applying for a position as a manager or administrator who oversees other financial examiners in an institution or organization. Financial examiners who start in state government or local organizations can also attempt to advance by accruing experience and applying for positions in corporations or in federal positions.

## EMPLOYMENT AND OUTLOOK

Financial examiners held 62,800 jobs in 2021. Demand for these workers in the financial industry increased because of the necessity for financial institutions to efficiently comply with federal regulation. More financial institutions are hiring financial examiners to help navigate the regulatory environment and reduce the cost of compliance. Financial examiners' employment is projected to grow 21 percent from 2021 to 2031 in the finance and insurance industry.

At the federal level, the creation of the Consumer Financial Protection Bureau (CFPB) has increased employment of financial examiners in recent years. However, changes to this agency and overall budget constraints in the federal government may limit employment growth.

Jobseekers with work experience in banking, insurance, or accounting, for example having worked as accountants and auditors or financial analysts, should have the best prospects.

### Related Occupations

- Accountants/Auditors
- Budget Analysts
- Financial Analysts

**Fast Fact**

**According to a report by professional services giant Deloitte, financial institutes dedicate 0.3% of revenue and 10% of their IT budget to cybersecurity.**

Source: stealthlabs.com

Photo via iStock.com/ipopba. [Used under license.]

- Loan Officers
- Management Analysts
- Personal Financial Advisors
- Private Detectives/Investigators
- Tax Examiners/Collectors/Revenue Agents

## MORE INFORMATION

**American Institution of Certified Public Accountants (AICPA)**
220 Leigh Farm Road
Durham, NC 27707-8110
888.777.7077
www.aicpa.org

**Association of Certified Fraud Examiners (ACFE)**
716 West Avenue
Austin, TX 78701-2727
800.245.3321
memberservices@acfe.com
www.acfe.com

**Consumer Financial Protection Bureau (CFPB)**
1700 G Street NW
Washington, DC 20552
855.411.2372
www.consumerfinance.gov

**Federal Deposit Insurance Corporation (FDIC)**
Washington, DC
877.275.3342
www.fdic.gov

**Society of Financial Examiners (SOFE)**
3505 Vernon Woods Drive
Summerfield, NC 27358
800.787.7633
info@sofe.org
www.sofe.org

---

*Emily Porter; updated by Daniel Vest*

## *Conversation With...*

## JIM KATTMAN, CFE

Managing Director, Noble Consulting Services, Inc.
Indianapolis, Indiana
Insurance regulation, 21 years

**What was your individual career path in terms of education/training, entry-level job, or other significant opportunity?**

In high school, I wasn't sure what I wanted to do but options in business seemed most intriguing, and accounting seemed a good entryway into the business world. I earned a bachelor's degree from Indiana University. I thought public accounting would give me exposure to different areas and when I went to work at a local Certified Public Accounting (CPA) firm in Indianapolis, it did. I worked on different types of accounts, and found insurance through a client. I moved over to Anthem Life Insurance Company of Indiana and did statutory reporting for a year before becoming an examiner performing statutory examinations for the Indiana Department of Insurance, which regulates all life and health, HMO, property and casualty insurers in Indiana. This includes nationally significant insurers.

Five years later, I took the opportunity to join Noble Consulting Services, Inc., which works or has worked with 15-20 state insurance departments in its 20-year history. But, in this specialized industry, we maintain relationships with most state insurance departments in the country.

States are responsible for regulating insurance companies. In my current role, I'm responsible for quality control for examination reports provided to state insurance departments. I am part of the team responsible for firm-wide controls and procedures related to examinations.

**What are the most important skills and/or qualities for someone in your profession?**

Being an insurance examiner requires technical knowledge and skill in applying laws and regulations to real-world situations, along with advanced written and oral communication ability.

It's important to be able to interact with different personalities and understand what people are saying because you're always talking with people who are experts in their different fields.

**What do you wish you had known going into this profession?**

The importance of building and maintaining relationships has time and again proven to be a valuable tool and resource. In a career such as this that requires technical skill and knowledge—and probably advanced degrees or designations—the pool of people pursuing the same or similar careers is likely to be limited in a regional or even national scope. That means you are likely to work with the same people and good relationships are critical.

**Are there many job opportunities in your profession? In what specific areas?**
Yes. Insurance is a highly specialized and regulated industry. The technical skills and knowledge required are in demand—especially in the accounting, finance, and legal areas—even as the insurance industry experiences consolidation in certain areas. In my field of expertise, there are in-house analyst functions and field-examination functions within state insurance departments. The roles within state insurance departments are diverse, meaning people with careers in many insurance-related fields may find themselves qualified to work in insurance regulation.

**How do you see your profession changing in the next five years, what role will technology play in those changes, and what skills will be required?**
Working remotely is rapidly increasing, which cuts down on the expense of travel. The ability to share information and documents electronically in real time has a real impact on what we do. Software improvements in functions such as customer relations management, or even keeping time and expense records, require technical knowledge of these programs and applications.

**What do you enjoy most about your job? What do you enjoy least about your job?**
I enjoy the opportunity to work with diverse and interesting people in many parts of the country. The relationships built over the years are important and rewarding and the perspective that comes from engaging with these people is interesting.

Working with people across the country means travel is often involved. Some travel is a great way to see parts of the country that you otherwise might not see. But over time, and as your life changes, travel can mean time away from family and friends. Everyone's situation is different and that can place travel requirements in a positive or negative light.

**Can you suggest a valuable "try this" for students considering a career in your profession?**
There may be intern opportunities with state insurance departments. Some independent consulting firms such as Noble offer intern opportunities throughout the year. This is a great way for students and recent graduates to better understand the work involved.

*This interview was originally published in 2017.*

# Financial Manager

## Snapshot

**Career Cluster(s):** Business Management & Administration; Finance

**Interests:** Accounting; Business; Finance; Data

**Earnings (Yearly Average):** $131,710

**Employment & Outlook:** Much Faster Than Average Growth Expected

## OVERVIEW

### Sphere of Work

Financial managers are responsible for the financial health of an organization. They create financial reports, direct investment activities, and develop plans for the long-term financial goals of their organization. Because of the highly sensitive financial information they deal in, financial managers must have a strong grasp of cybersecurity risks and constantly work at implementing new controls to address the growing threat to the organizations they supervise.

### Work Environment

Financial managers work in many industries, including banks, investment firms, and insurance companies. Most financial managers work full time, and some work more than 40 hours per week.

Photo via iStock.com/FG Trade. [Used under license.]

### Occupation Interest

Those individuals interested in pursuing a career path in Financial Management will be necessarily business-oriented, with an enterprising and social nature. These people tend to be leaders who not only excel when it comes to dealing with numbers, but people too. Their work will frequently involve working with, communicating with, and teaching people. As one of the main goals of financial management is to maintain financial health of those they work for, they must have an interest in helping or providing service to others. Part of doing so includes an awareness of cybersecurity interests, and protecting sensitive information.

### *Profile*

**Working Conditions:** Inside
**Education Needs:** Bachelor's Degree
**Licensure/Certification:** Not Required
**Opportunities for Experience:** Prior experience in related occupations
**Interest Score:** ECS

### *Duties and Responsibilities*

- Preparing financial statements, business activity reports, and forecasts
- Monitoring financial details to ensure that legal requirements are met
- Supervising employees who do financial reporting and budgeting
- Reviewing financial reports and seeking ways to reduce costs
- Analyzing market trends to maximize profits and finding expansion opportunities
- Helping management make financial decisions

### A Day in the Life—Duties and Responsibilities

Financial managers spend much of their time analyzing data and advising senior managers on ways to maximize profits. They often work on teams, acting as advisors to top executives.

Financial managers must have knowledge of the topics, tax laws, and regulations that are specific to their organization or industry. For example, government financial managers must be experts on appropriations and budgeting processes; healthcare financial managers must understand billing, reimbursement, and other business matters related to healthcare.

## OCCUPATION SPECIALTIES

### Cash Manager

Cash managers monitor and control the flow of money into and out of an organization to meet business and investment needs. For example, they must project whether the organization will have a shortage or surplus of cash.

### Controller

Controllers direct the preparation of financial reports that summarize and forecast an organization's financial position. These reports may include income statements, balance sheets, and analyses of future earnings or expenses. Controllers also are in charge of preparing reports required by governmental agencies that regulate businesses. Often, controllers oversee the accounting, audit, and budget departments of their organization.

**Credit Manager**

Credit managers oversee an organization's credit business. They set credit-rating standards, determine credit limits, and monitor the collections of past-due accounts.

**Insurance Manager**

Insurance managers decide how to limit an organization's losses by protecting against risks, such as for disability payments to an employee who gets hurt on the job or for costs imposed by a lawsuit against the organization.

**Risk Manager**

Risk managers use strategies to limit or offset an organization's chance of financial loss or exposure to financial uncertainty. Among the risks they try to limit are those arising from currency or commodity price changes.

**Treasurer/Finance Officer**

Treasurers and finance officers direct an organization's budgets to meet its financial goals. They oversee investments and other plans to raise capital, such as issuing stocks or bonds, to support their organization's growth. They also develop financial plans for mergers (two companies joining together) and acquisitions (one company buying another).

## WORK ENVIRONMENT

### Immediate Physical Environment

Financial managers will typically work in offices, sometimes attending conferences and meetings boardroom settings.

### Human Environment

Financial managers work closely with top executives and with departments that develop data needed for analysis. Communication skills are one of the top requirements of the job for this reason: effectively communicating ideas in oral and written reports is essential.

### Technological Environment

A high degree of computer literacy will be required in this position. Familiarity with various software programs, from accounting software (such as QuickBooks), to business intelligence and data analysis software (e.g. Cognos Impromptu; MicroStrategy), to ERP software, may be necessary.

## EDUCATION AND TRAINING

### High School/Secondary

High school students interested in becoming financial managers should focus on math, statistics, and business-related courses. Entering this profession will often require at least a bachelor's degree, so students should prepare themselves for postsecondary education, including researching the schools they would like to consider attending to achieve that goal.

**Suggested High School Subjects**

- Algebra
- Economics
- Pre-Calculus
- Psychology
- Statistics
- Trigonometry

**Related Career Pathways/Majors**

*Business Management & Administration Career Cluster*

- General Management Pathway
- Operations Management Pathway

*Finance Career Cluster*

- Accounting Pathway
- Banking Services Pathway
- Business Finance Pathway
- Securities & Investments Pathway

**Transferable Skills and Abilities**

**Analytical Skills**

- Evaluating data and information that affects their organization

**Communication Skills**

- Explaining and justifying complex financial transactions

**Detail-oriented**

- Paying attention to details when compiling and examining documents

**Math Skills**

- Analyzing, comparing, and interpreting facts and figures
- Understanding international finance and complex financial documents
- Using advanced math skills, such as calculus and statistical analysis

**Organizational Skills**

- Having structures in place to be effective in their work

### Postsecondary

Financial managers typically need a bachelor's degree and 5 years or more of experience in another business or financial occupation, such as an accountant, securities sales agent, or financial analyst.

Financial managers typically need at least a bachelor's degree in business, economics, or a related field. These disciplines help students learn analytical skills and methods.

**Related College Majors**

- Accounting
- Auditing
- Business Management & Administration
- Economics
- Entrepreneurial Studies
- Financial Management
- General Business
- Insurance
- International Business
- Marketing

## Adult Job Seekers

Adults seeking to become financial managers should have a background or strength in math or finance. Financial managers usually have experience in another business or financial occupation. For example, they may have worked as a loan officer, accountant, securities sales agent, or financial analyst.

In some cases, companies provide management training to help prepare motivated, skilled financial workers to become managers.

## Professional Certification and Licensure

Although it is not required, professional certification indicates competence for financial managers who have it. The Association of Government Accountants (AGA) offers the Certified Government Financial Manager (CGFM) designation to financial managers working with federal, state, or local government. To earn this certification, candidates must have a bachelor's degree from an accredited college or university, pass examinations, and have professional-level experience in government financial management. To keep the certification, CGFMs must complete continuing professional education.

The CFA Institute confers the Chartered Financial Analyst (CFA) certification to investment professionals who have at least a bachelor's degree or 4 years of work experience, or a combination of experience and education, and who pass three exams. The Association for Financial Professionals confers the Certified Treasury Professional (CTP) credential to those who have at least 2 years of relevant experience or 1 year of experience and a graduate degree in business, finance, or a related field. This association also confers the Certified Corporate Financial Planning Analysis Professional (FP&A) credential to those who have a bachelor's degree or who are currently enrolled in an undergraduate program with a finance-related major and will graduate within 2 years. Both credentials require passing an exam.

Certified public accountants (CPAs) are licensed by their state's board of accountancy and must pass an exam administered by the American Institute of Certified Public Accountants (AICPA).

## EARNINGS AND ADVANCEMENT

The median annual wage for financial managers was $131,710 in 2021. The lowest 10 percent earned less than $77,040, and the highest 10 percent earned more than $208,000.

Experienced financial managers may advance to become chief financial officers (CFOs). These executives are responsible for the accuracy of an organization's financial reporting.

## EMPLOYMENT AND OUTLOOK

Employment of financial managers is projected to grow 17 percent from 2021 to 2031, much faster than the average for all occupations.

About 71,300 openings for financial managers are projected each year, on average, over the decade. Many of those openings are expected to result from the need to replace workers who transfer to different occupations or exit the labor force, such as to retire.

Services provided by financial managers, such as planning, directing, and coordinating investments, are likely to stay in demand as the economy grows. In addition, several specialties within financial management, particularly cash management and risk management, are expected to be in high demand over the decade.

Companies, particularly those with operations in foreign countries, have accumulated more cash on their balance sheets in recent years. As globalization contin-

## Famous First

**The WEF's "Global Security Outlook 2022" found that "only 19% of cyber leaders feel confident that their organisations are cyber resilient." To prevent and mitigate cyber risk, finance team leaders are encouraged not to necessarily throw money at the flashiest security technologies, but instead to remember that the leading cause of data breaches is human error; proper training and vigilance will be of the greatest benefit.**

Source: Financial Management Magazine

Photo via iStock.com/nd3000. [Used under license.]

ues, this trend is likely to persist. This practice should lead to demand for financial managers, as companies will need expertise in managing cash.

There has been an increased emphasis on risk management within the financial industry, and this trend is expected to continue. Banking institutions are expected to emphasize stability and managing risk over profits. This emphasis is expected to lead to employment growth for risk managers.

### Related Occupations

- Accountant/Auditor
- Budget Analyst
- Financial Analyst
- Insurance Sales Agent
- Insurance Underwriter
- Loan Officer
- Personal Financial Advisor

## MORE INFORMATION

**American Institute of Certified Public Accountants (AICPA)**
1345 Avenue of the Americas, 27th Floor
New York, NY 10105
212.596.6200
www.aicpa.org

**Association for Financial Professionals (AFP)**
4520 East-West Highway, Suite 800
Bethesda, MD 20814
301.907.2862
customerservice@AFPonline.org
www.afponline.org

**Association of Government Accountants (AGA)**
2208 Mount Vernon Avenue
Alexandria, VA 22301-1314
703.684.6931
www.agacgfm.org

**CFA Institute**
1401 New York Avenue NW, Suite 330
Washington, DC 20005
434.951.5499
info@cfainstitute.org
www.cfainstitute.org

**Global Academy of Finance and Management (GAFM)**
P.O. Box 293
1670F East Cheyenne Mountain Boulevard
Colorado Springs, CO 80906
www.gafm.com

---

*Daniel Vest*

## *Conversation With...*
# JOSEPH NATIVO

Kumon, Teaneck, NJ
CFO, 26 years

**What was your individual career path in terms of education/training, entry-level job, or other significant opportunity?**

I worked as an apprentice for a small CPA firm during college. I learned how to prepare bank reconciliations, payroll tax returns, and income tax returns. Back then, we prepared tax returns by hand. Looking back, preparing the returns manually was the best experience I could have ever received. It gave me a solid foundation and made the eventual switch to tax software a cinch. After graduating from Rutgers University with a bachelor's in Accounting and earning my CPA license, I spent several years working for a mid-sized firm. Although I enjoyed working in public accounting tremendously, I wanted to experience the private industry. I accepted an accounting manager position at Kumon North America Inc., a franchisor of after-school math and reading centers. The U.S. operations were quite small when I joined the company back in 1992. Today, Kumon is the market leader with nearly 1,500 locations and over 275,000 students in the U.S.

**What are the most important skills and/or qualities for someone in your profession?**

Today's CFOs are expected to wear multiple hats. You have to be adaptable and exhibit a willingness to take on new challenges. Having a thorough understanding of the business operations is just as important as knowing the financial side because with this knowledge, the CFO can support executive management in strategic planning. Strong interpersonal skills and the ability to manage at all levels are paramount. A CFO must possess strong problem-solving skills and make decisions based on accurate and timely analyses. Finally, a high level of integrity, a strong sense of urgency, and being results-oriented are core qualities of a CFO.

**What do you wish you had known going into this profession?**

Unique corporate cultures exist at every organization. Learning how to work within the culture isn't something taught while attending university. Strong interpersonal skills and adaptability come with time and experience. Therefore, don't be afraid to ask questions. Pick the brains of others in your organization and absorb as much as possible.

**Are there many job opportunities in your profession? In what specific areas?**
Demand for accounting positions is at an all-time high. According to the U.S. Bureau of Labor Statistics, a growth rate of 13 percent is projected through 2022. The unemployment rate for accountants is much lower when compared to the national average. *U.S. News and World Report's* 2014 Best Jobs issue ranked Accounting as the 3rd best career.

Positions are available in public accounting, private industry, government, and not-for-profit organizations. Many higher level positions will require a CPA or MBA as well.

**How do you see your profession changing in the next five years? What role will technology play in those changes, and what skills will be required?**
Cloud, mobile, social media, and Big Data are the emerging technologies with the most impact on the accounting profession, according to a survey of more than 400 U.S. accountants released by software provider CCH, a Wolters Kluwer business. These technologies allow professionals to provide financial data more effectively to end-users. As the automation of data collection continues to increase, accounting professionals will be able to shift focus from crunching numbers to analyzing data and advising key stakeholders

**What do you like most about your job? What do you like least about your job?**
I really believe in Kumon's corporate mission. We help children unlock their potential through self-learning with the aim to foster sound, capable individuals who can contribute to the global community. I've always taken pride in the fact that we make a difference in the lives of children worldwide. Also, my role is not limited to debits and credits. I like being involved with different areas of the operations and enjoy being a resource for my colleagues. If I can offer information that allows them to perform better and contribute to company goals, then I will continue to find my work meaningful and fruitful.

As in most professions, the hours can be long and the pressure can be high. Therefore, work-life balance is very important. I always remind my staff to use their paid vacation time whenever possible. We have to allow ourselves to recharge our batteries from time to time.

**Can you suggest a valuable "try this" for students considering a career in your profession?**
Seek an internship as early as possible. Don't wait until your senior year. The point is to get your feet wet and gain real world experience. The knowledge acquired will be invaluable after you graduate. Hopefully, an internship will lead to a full-time position. Regardless, any experience you gain will shed insight on how public or private firms operate. It will also help you figure out which path to seek in the business world. An accounting degree with a CPA and/or MBA will open many doors.

*This interview was originally published in 2015.*

# Forensic Science Technician

### Snapshot

**Career Cluster(s):** Law, Public Safety, Corrections & Security; Science, Technology, Engineering & Mathematics

**Interests:** Data; Medicine; Science; Statistics; Investigating

**Earnings (Yearly Average):** $61,930

**Employment & Outlook:** Much Faster Than Average Growth Expected

## OVERVIEW

### Sphere of Work

Forensic science technicians aid criminal investigations by collecting and analyzing evidence. Digital forensics involves the gathering, analyzation and securing of digital evidence for forensic investigative purposes.

Photo via iStock.com/sanjeri. [Used under license.]

### Work Environment

Most laboratory forensic science technicians work during regular business hours. Crime scene investigators may work extended or unusual hours and travel to crime scenes within their jurisdiction.

### Occupation Interest

Forensic science technicians possess an interest in contributing to law enforcement and participating in criminal investigations, using their knowledge to secure evidence of criminal wrongdoing.

### Profile

**Working Conditions:** Varies
**Education Needs:** Bachelor's Degree
**Licensure/Certification:** Various; Not usually required
**Opportunities for Experience:** On-the-job training
**Interest Score:** IRC

## A Day in the Life—Duties and Responsibilities

Forensic science technicians aid criminal investigations by collecting and analyzing evidence. Many technicians specialize in either crime scene investigation or laboratory analysis. Forensic science technicians may be generalists who perform many duties, or they may specialize in certain techniques and sciences. Generalist forensic science technicians, sometimes called criminalists or crime scene investigators, collect evidence at the scene of a crime and perform scientific and technical analysis in laboratories or offices.

Forensic science technicians who work primarily in laboratories may specialize in the natural sciences or engineering. These workers, such as forensic biologists and forensic chemists, typically use chemicals and laboratory equipment such as microscopes when analyzing evidence. They also may use computers to examine DNA, substances, and other evidence collected at crime scenes. They often work to match evidence to people or other known elements, such as vehicles or weapons. Most forensic science technicians who perform laboratory analysis specialize in a specific type of evidence, such as DNA or ballistics.

### Duties and Responsibilities

- Analyzing crime scenes to determine what evidence should be collected and how
- Taking photographs of the crime scene and evidence
- Making sketches of the crime scene
- Recording observations and findings, such as the location and position of evidence
- Collecting evidence, including weapons, fingerprints, and bodily fluids
- Cataloging and preserving evidence for transfer to crime labs
- Reconstructing crime scenes
- Performing chemical, biological, and microscopic analyses on evidence taken from crime scenes
- Exploring possible links between suspects and criminal activity, using the results of DNA or other scientific analyses
- Consulting with experts in specialized fields, such as toxicology (the study of poisons and their effect on the body) and odontology (a branch of forensic medicine that concentrates on teeth)

Some forensic science technicians, called forensic computer examiners or digital forensics analysts, specialize in computer-based crimes. They collect and analyze data to uncover and prosecute electronic fraud, scams, and identity theft. The abundance of digital data helps them solve crimes in the physical world as well. Computer forensics technicians must adhere to the same strict standards of evidence gathering found in general forensic science because legal cases depend on the integrity of evidence.

## WORK ENVIRONMENT

### Immediate Physical Environment

Many if not all forensic science technicians work in laboratory environments; but some, known as criminalists or crime scene investigators, collect evidence at the scene of a crime, thus their work could take them anywhere be it indoors or outdoors, and expose them to potentially grisly scenes.

### Human Environment

Forensic science technicians must work in collaboration with other scientists as well as law enforcement officials on a regular basis. There is a marked need to communicate in clear and effective language sometimes complex scientific notions, so that non-technical persons—such as officers, or jurors—can understand the evidence being presented.

### Technological Environment

Forensic science technicians work with a variety of technology in laboratory environments, studying evidence, applying scientific method and reconstructing crime scenes. These professionals, especially in the realm of dealing with cybercrime, may possess a vast knowledge of various computer components and how to apply digital forensics techniques and tools on computers and other wired and mobile devices to investigate. They may also be expected to handle information extraction, as well as how to identify information that is useful in a legal case and how to explain and present it in court.

## EDUCATION AND TRAINING

### High School/Secondary

High school students interested in becoming forensic science technicians should focus on math and science-related courses. Those interested in digital forensics specifically should also include courses such as related to computer learning. This profession requires at least a bachelor's degree, so students should prepare themselves for postsecondary education, including researching the schools they would like to consider attending to achieve that goal.

**Suggested High School Subjects**

- Algebra
- Biology
- Chemistry
- Computer Programming
- Earth or Life or Physical Science
- English
- Geography
- Geometry
- Physics
- Pre-Calculus
- Psychology
- Statistics
- Trigonometry

**Related Career Pathways/Majors**

*Health Science Career Cluster*

- Biotechnology Research & Development Pathway

*Law, Public Safety, Corrections & Security Career Cluster*

- Law Enforcement Services Pathway

*Science, Technology, Engineering & Mathematics Career Cluster*

- Science & Mathematics Pathway

### Transferable Skills and Abilities

**Communication Skills**

- Writing reports and testifying in court
- Working with other law enforcement officials and specialists

**Critical-thinking Skills**

- Using their best judgment when matching physical evidence, such as fingerprints and DNA, to suspects

**Detail-oriented**

- Noticing small changes in mundane objects
- Being good at collecting and analyzing evidence
- Paying attention to details when compiling and examining documents

**Math Skills**

- Analyzing, comparing, and interpreting facts and figures
- Using advanced math skills, such as calculus and statistical analysis

**Organizational Skills**

- Working with a range of financial documents for a variety of clients

## Postsecondary

Forensic science technicians typically need at least a bachelor's degree. On-the-job training is usually required both for those who investigate crime scenes and for those who work in labs.

Forensic science technicians typically need at least a bachelor's degree in a field such as physical science, biology, or forensic science. Forensic science programs may specialize in a specific area of study, such as toxicology, pathology, or DNA. Students who enroll in general natural science programs should make an effort to take classes related to forensic science. A list of schools that offer degrees in forensic science is available from the American Academy of Forensic Sciences. Many of those who seek to become forensic science technicians will have an undergraduate degree in the natural sciences and a master's degree in forensic science.

## Famous First

**The idea of using fingerprints as a form of identification was first developed in the seventh century. According to historical accounts, a street vendor once used a customer's fingerprints to identify a specific person who owed him money. The prints were approved by a judge as indisputable proof, offering a very early illustration of forensic science in action.**

Source: crimemuseum.org

Photo via iStock.com/Michael Burrell. [Used under license.]

Many crime scene investigators who work for police departments are sworn police officers and have met educational requirements necessary for admittance into a police academy. Applicants for civilian crime scene investigator jobs should have a bachelor's degree in either forensic science, with a strong basic science background, or the natural sciences. For more information on police officers, see the profile on police and detectives.

Forensic science technicians receive on-the-job training before they are ready to work on cases independently. Newly hired crime scene investigators may work under experienced investigators while they learn proper procedures and methods for collecting and documenting evidence.

Forensic science technicians learn laboratory specialties on the job. The length of this training varies by specialty, but is usually less than a year. Technicians may need to pass a proficiency exam or otherwise be approved by a laboratory or accrediting body before they are allowed to perform independent casework.

Throughout their careers, forensic science technicians need to keep up with advances in technology and science that improve the collection or analysis of evidence.

**Related College Majors**

- Accounting
- Auditing
- Business Management & Administration
- Economics
- Entrepreneurial Studies
- Financial Management
- General Business

- Insurance
- International Business
- Marketing

### Professional Certification and Licensure

A range of licenses and certifications is available to help credential, and aid in the professional development of, many types of forensic science technicians. Certifications and licenses are not typically necessary for entry into the occupation. Credentials can vary widely because standards and regulations vary considerably from one jurisdiction to another.

## EARNINGS AND ADVANCEMENT

The median annual wage for forensic science technicians was $61,930 in 2021. The lowest 10 percent earned less than $37,670, and the highest 10 percent earned more than $103,430.

## EMPLOYMENT AND OUTLOOK

Forensic science technicians held about 17,600 jobs in 2021. Employment is projected to grow 11 percent from 2021 to 2031, much faster than the average for all occupations.

About 2,500 openings for forensic science technicians are projected each year, on average, over the decade. Many of those openings are expected to result from the need to replace workers who transfer to different occupations or exit the labor force, such as to retire.

Because this is a small occupation, its fast growth is expected to result in only about 2,000 new jobs over the projections decade.

State and local governments are expected to hire additional forensic science technicians to process their high caseloads. Additionally, scientific and technological advances are expected to increase the availability, reliability, and usefulness of objective forensic information used as evidence in trials. As a result, more forensic science technicians will be needed to provide forensics information to law enforcement agencies and courts.

### Related Occupations

- Biological Technician
- Chemical Technician
- Chemist/Materials Scientist
- Clinical Laboratory Technologist/Technician
- Environmental Science/Protection Technician
- Fire Inspector
- Hazardous Materials Removal Worker
- Police/Detective
- Private Detective/Investigator

## MORE INFORMATION

**American Academy of Forensic Sciences (AAFS)**
410 North 21st Street
Colorado Springs, CO 80904
719.636.1100
info@aafs.org
www.aafs.org

**American Board of Criminalistics**
P.O. Box 1358
Palmetto, FL 34220
941.729.9050
acbcregistrar@criminalistics.com
www.criminalistics.com

**American Board of Medicolegal Death Investigators (ABMDI)**
10104 Senate Drive, Suite 241
Lanham, MD 20706
410.807.3007
info@abmdi.org
www.abmdi.org

**Association of Firearm and Tool Mark Examiners (AFTE)**
550 West 16th Street
Indianapolis, IN 46202
webmaster@afte.org
www.afte.org

**International Crime Scene Investigators Association (ICSIA)**
P.O. Box 1532
Grand Rapids, MI 49501
800.504.7501
icsia@icsia.org
www.icsia.org

---

*Daniel Vest*

## *Conversation With...*

## BARRY A. J. FISHER

Crime Laboratory Director, Retired
LA County Sheriff's Department, Los Angeles, California
In the field, 50 years

**What was your individual career path in terms of education/training, entry-level job, or other significant opportunity?**

I received an MS degree in chemistry and accidentally discovered forensic science while searching for my first job following graduate school. After starting work at the crime lab, I also pursued a masters degree in business administration in an evening program while at the laboratory. I suspect the MBA helped me to promote into the supervisory and later management ranks of the crime lab.

**What are the most important skills and/or qualities for someone in your profession?**

The most important skills are a strong background in science—chemistry or biological sciences—and good communication skills. The most important qualities are a strong sense of professional ethics and the ability to work within bureaucratic systems, especially if you are working in government organizations.

**What do you wish you had known going into this profession?**

Working in a law enforcement environment will cause you to see the world very differently. You see the worst of humanity and must be careful not to project that notion to all people you encounter.

**Are there many job opportunities in your profession? In what specific areas?**

There are many job opportunities in forensic sciences but there is lots of competition. Be prepared to market yourself for any position. Often graduate degrees are important. The current major areas in forensic science are in: DNA, drug and toxicology, digital evidence (computer forensics), and forensic chemistry.

**How do you see your profession changing in the next five years, how will technology shift, and what skills will be required?**

Automation and artificial intelligence (AI) are likely to become more prevalent in crime labs. As a result, computer-related skills may become more valuable.

### What do you enjoy most about your job? What do you enjoy least about your job?

I enjoyed the variability of the work. No two days were ever the same and if you enjoy solving problems, this is the career for you.

The least enjoyable part is the constant exposure to criminal acts. You start to become less forgiving of those who commit crimes. This can rub off onto your view of society.

### Can you suggest a valuable "try this" for students considering a career in your profession?

If possible, aspirants should seek out internship opportunities. Those with public speaking fears should join Toastmasters International to hone their speaking abilities. Graduate degrees are more sought after by employers. Visit crime laboratories and ask for a tour. Attend a regional or national forensic science meeting. Apply for as many job opportunities as you can. It's an easy way to practice your interview skills.

*This interview was originally published in 2021.*

## *Conversation With...*

# CARLA MILLER NOZIGLIA, DISTINGUISHED FELLOW (AAFS)

Forensic Scientist, retired
Senior Forensic Advisor, Aiken, South Carolina
In the field, 44 years

**What was your individual career path in terms of education/training, entry-level job, or other significant opportunity?**

Since forensic science was virtually unknown when I went to school, there were no degrees nor formal training programs. (Fingerprints were done, but by officers, some of whom were injured, and not able to work on the street, so were sent to "the lab.") So, my 1963 BS is in Medical Technology/Biology, minor in Chemistry. I was assigned to special chemistry in a hospital laboratory where we did marijuana testing for law enforcement. And I was hooked. I learned on the job, and at the bench. I later got my MS in Scientific Administration.

**What are the most important skills and/or qualities for someone in your profession?**

Most important quality is honesty. Most important skill is curiosity. The test result is not always what we, or law enforcement, want it to be. Truth is more important than a conviction. Curiosity will take you from the mundane test(s) that is requested, to the truth, which may entail more testing.

**What do you wish you had known going into this profession?**

I had no pre-conceived ideas because it was all so new. I never knew my career would be so interesting.

I have edited two editions of *The Forensic Laboratory Handbook* and written the introduction to a series for teens, *The Crime Scene Club, Fact and Fiction.*

Most interesting is the decade I spent in Tanzania as the Senior Forensic Advisor for the International Criminal Investigative Training Assistance Program (ICITAP), of the US Department of Justice. My mission was to bring the Tanzania Police Force Laboratory up to current standards with instrumentation, training, and testing.

**Are there many job opportunities in your profession? In what specific areas?**
Forensic science has taken off so much so that a case with no forensic evidence is not what a jury wants to hear. From fingerprints, there are now so many forensic disciplines: criminalistics, anthropology, engineering, firearms, nursing, psychology/behavioral science, toxicology, accounting, pathology, and digital/multimedia science, to name a few. There are many jobs in public, private, and government areas in known disciplines and others which will surely develop.

**How do you see your profession changing in the next five years, how will technology shift, and what skills will be required?**
As with most fields, automation and digital applications will expand, giving results quicker, answers more in-depth, and, hopefully, more accuracy. Basic skills of observation; small motor skills to process evidence; ability to transfer scientific results to a report understood by a nonscientist; and the ability to testify to a jury understandably will still be essential.

**What do you enjoy most about your job? What do you enjoy least about your job?**
I loved almost everything when I worked cases: solving a mystery, finding what couldn't be seen, being part of the justice system. It is just as important to incarcerate the bad guy as to eliminate the good guy. I loved to testify, seeing the flash of understanding as I was explaining my tests.

As I moved up to administration, I learned that scientists who have been taught to work and think independently are not the easiest employees to manage.

**Can you suggest a valuable "try this" for students considering a career in your profession?**
Because of confidentiality, it is difficult to "tour" a forensic laboratory. However, internships are available; universities offer forensic courses and public speaking courses.

*This interview was originally published in 2021.*

# *Conversation With...*

## JOY REHO

Case Management Supervisor, Forensic Lab
29 years in field

**What was your individual career path in terms of education/training, entry-level job, or other significant opportunity?**

When I went to college initially, I majored in liberal arts because I didn't know what I was going to pursue. After two years, I realized I enjoyed science. I majored in liberal arts and sciences with a concentration in biological sciences. After I received my bachelor's degree in biological sciences, I applied to and was accepted to a master's program in forensic science. My first job was at the state of Connecticut forensic lab in Forensic Serology, which is the study of blood and body fluids. After two years, I was promoted to Criminalist and two years after that, I was promoted to Lead Criminalist. For the last 24 years, I was a Supervisor in Forensic Biology. I recently became a supervisor in case management, which involves prioritizing cases, streamlining analyses and coordinating the workflow in the lab.

**What are the most important skills and/or qualities for someone in your profession?**

To be a successful criminalist, you must be very detail-oriented and have excellent organizational skills. You have to be a careful and accurate worker, because our work is scrutinized by attorneys in court. The results of our testing may exonerate someone or be the deciding factor in a conviction. The heavy case load necessitates a strong work ethic and the ability to work independently.

One misconception about forensic science is how quickly crimes can be solved. It takes more than a few hours to complete a case. This is dubbed the "CSI Effect." Popular TV shows depict lab personnel that are experts in every area. Typical forensic scientists are experts in one area of specialty, for example firearms, fingerprints, trace analysis, DNA, or blood and body fluid identification. No one is like Abby on *NCIS* or the scientists on *Bones* or *CSI*. Also, juries expect to see evidence examined in a crime lab, so even if there is a confession, the evidence needs to be tested.

**What do you wish you had known going into this profession?**

I wish I had known when I was still in school how much I would have to rely on strong math skills in my day-to-day job. You have to have a solid footing in statistics and molecular genetics. Additionally, I wish I had stronger computer skills when I started out.

**Are there many job opportunities in your profession? In what specific areas?**

Yes, there are. One of the fields within forensic science that is seeing the most growth involves computer crimes. The field of molecular biology—and DNA in particular—is also burgeoning.

**How do you see your profession changing in the next five years? What role will technology play in those changes, and what skills will be required?**

Technology will definitely play a role in the changes that are coming. There will be more emphasis on robotics for DNA extraction because it allows a faster throughput of sampling and extraction. Additionally, newer techniques are more sensitive, allowing for a smaller sample size. Finally, computer forensics will only continue to become more commonplace and important in the next five or more years.

**What do you enjoy most about your job? What do you enjoy least?**

What I enjoy most about my job is the fulfillment I get from assisting police departments in solving crimes. It's a feeling of accomplishment when I use science to bring closure to victims. What I enjoy least is the volume of casework. It can really be overwhelming at times. There's a constant struggle for funding to obtain supplies and equipment that we need to do our job. This can be frustrating and time-consuming.

**Can you suggest a valuable "try this" for students considering a career in your profession?**

Students considering working as a criminalist could undertake a research project on an area of interest in forensic science. Are you interested in fingerprinting techniques? What about animal hair identification? Researching topics like this will help you discover how engaging you find the subject, and that, in turn, may help you figure out if this is really something you'd want to do for a career. You could prepare a display for a science fair and volunteer to show it at a Girl Scout or Boy Scout meeting. Are you able to clearly explain the topic?

Our lab has internships for college students. However, due to the nature of what we do, job shadowing is not an option. A major concern is contamination of the evidence by anyone near the evidence. Anyone entering the lab needs to provide a DNA sample that will be entered into the staff index of DNA profiles.

*This interview was originally published in 2019.*

# Human Resources Specialist

**Snapshot**

**Career Cluster(s):** Business Management & Administration

**Interests:** Administration; Office Work; Interviewing; Personnel Matters; Training

**Earnings (Yearly Average):** $62,290

**Employment & Outlook:** Faster Than Average Growth Expected

## OVERVIEW

### Sphere of Work

Human resources (HR) specialists recruit, screen, and interview job applicants and place newly hired workers in jobs. They also may handle compensation and benefits, training, and employee relations. HR can serve as a liaison between the information technology (IT) security department and the personnel at their companies, clarifying cybersecurity policy, giving resources, and working behind the scenes to identify and foresee potential information security concerns that may arise.

Photo via iStock.com/Harbucks. [Used under license.]

### Work Environment

Some organizations contract recruitment and placement work to outside firms, such as those in the employment services industry or the professional, scientific, and technical industry.

Human resources specialists generally work in office settings. Some, particularly recruitment specialists, travel to attend job fairs, visit college campuses, and meet with applicants.

Most specialists work full time during regular business hours. Some work more than 40 hours per week.

### Profile

**Working Conditions:** Inside
**Education Needs:** Bachelor's Degree
**Licensure/Certification:** Various
**Opportunities for Experience:** Volunteering; Internship
**Interest Score:** ECS

### Occupation Interest

Some of the traits that make for a good human resources specialist include: organized, effective communicators, good at solving problems and comfortable with making hard decisions. They are guided by strong ethics and display kindness and professionalism in dealing with sometimes sensitive issues in the workplace.

### Duties and Responsibilities

- Consulting with employers to identify hiring needs
- Interviewing job applicants about their relevant experience, education, and skills
- Checking applicants' references and backgrounds
- Informing applicants about job details, such as duties, benefits, and working conditions
- Hiring or referring qualified applicants
- Running or helping with new employee orientation
- Keeping employment records and processing paperwork

### A Day in the Life—Duties and Responsibilities

Human resources specialists often are trained in tasks for all disciplines of a human resources department. In addition to recruiting applicants and placing workers, human resources specialists help guide employees through human resources procedures and answer questions about an organization's policies. They sometimes administer benefits, process payroll, and handle associated questions or problems. Some specialists focus more on strategic planning and hiring than on administrative duties. They also ensure that all human resources functions comply with federal, state, and local regulations.

## OCCUPATION SPECIALTIES

### Human Resource Generalist

Human resources generalists handle all aspects of human resources work. Their duties include recruitment, compensation, benefits, training, and employee relations, as well as administering human resources policies, procedures, and programs.

**Recruitment Specialist/Recruiter/Talent Acquisition Specialist**
Recruitment specialists, sometimes known as recruiters or "talent acquisition specialists," find, screen, and interview applicants for job openings in an organization. They search for applicants by posting listings, attending job fairs, and visiting college campuses. They also may test applicants, contact references, and extend job offers.

## WORK ENVIRONMENT

### Immediate Physical Environment

Human resources specialists generally work in office settings, but travel may be required, particularly in the role of recruitment specialist. These workers travel to attend job fairs, visit college campuses, and meet with applicants.

### Human Environment

Interviewing job applicants and consulting with management are among the primary duties of human resources specialists. A sociable nature and facility with words is a must-have for anyone seeking the position, as the job is very much about how people relate to one another. Human resources specialists will have to deal in sensitive matters and have a good deal of tact, such as when dealing with employee relations issues, such as harassment allegations, work complaints, or other employee concerns.

### Technological Environment

Technology used by human resources specialists includes various software program types including specialized HR software, the purpose of which is to better allocate their time and resources to more productive, profitable efforts.

## EDUCATION AND TRAINING

### High School/Secondary

High school students interested in becoming human resource specialists should focus on business-related courses and courses such as English that focus on how people communicate and relate to one another. This job often requires a bachelor's degree, so students should prepare themselves for postsecondary education, including researching the schools they would like to consider attending to achieve that goal.

**Suggested High School Subjects**

- Civics
- Communications
- Computer Applications
- English
- Gender Studies
- Introduction to Business
- Psychology
- Sociology

**Related Career Pathways/Majors**

*Business Management & Administration Career Cluster*

- General Management Pathway
- Human Resources Management Pathway

### Transferable Skills and Abilities

**Communication Skills**

- Conveying information effectively and respond to questions and concerns from employers, job applicants, and employees

**Decision-making Skills**

- Using sound judgment when reviewing applicants' qualifications or when working to resolve disputes

**Detail-oriented**

- Paying attention to detail when evaluating applicants' qualifications, doing background checks, maintaining records of an employee grievance, and ensuring that a workplace complies with labor standards

**Interpersonal Skills**

- Continually interacting with others
- Conversing and connecting with people from varied backgrounds

## Postsecondary

Human resources specialists typically need a bachelor's degree in human resources, business, communications, or a related field.

By working in an internship during college, students gain relevant experience that may be helpful in competing for human resources specialist jobs. Internships in human resources departments may help prospective specialists to increase their understanding of the occupation and to network in an industry.

Some positions require human resources specialists to have relevant work experience. Candidates may gain experience as human resources assistants (information clerks), customer service representatives, or in related occupations.

Employers also may prefer to hire candidates who have experience in areas such as personnel recruitment, staff training and development, employee relations, and compensation and benefits. Candidates sometimes get this experience while in college, either through courses or by volunteering.

**Related College Majors**

- Business Management & Administration
- Communications
- General Business
- Human Resources
- International Business

## Fast Fact

**Increasingly, HR departments use chatbots for recruitment, benefits process automation and employee interaction. However, this comes with security risks: if employees send sensitive information through chatbot channels, the information could subsequently be exposed via web browser caches and server log files, or even exploited via malware. Employees should exercise caution when it comes to the data they send, to prevent future problems.**

Source: techtarget.com

Photo via iStock.com/Tero Vesalainen. [Used under license.]

### Adult Job Seekers

Adults seeking to become human resources specialists should have a background or strength in business, if not a bachelor's degree human resources, communications, or a related field.

### Professional Certification and Licensure

Professional associations that specialize in human resources offer courses to enhance the skills of their members, and some offer certification programs. For example, the Society for Human Resource Management (SHRM) offers the SHRM Certified Professional (SHRM-CP) and SHRM Senior Certified Professional (SHRM-SCP), and the HR Certification Institute (HRCI) offers a range of certifications for varying levels of expertise.

Certification usually requires that candidates pass an exam that covers human resources knowledge and asks candidates to apply their knowledge to different situations. Candidates for certification also typically need to meet minimum education and experience requirements.

Although certification is usually voluntary, some employers prefer or require it. Human resources generalists, in particular, may benefit from certification because it shows knowledge and professional competence across all human resources areas.

## EARNINGS AND ADVANCEMENT

The median annual wage for human resources specialists was $62,290 in 2021. The lowest 10 percent earned less than $37,680, and the highest 10 percent earned more than $108,160.

Some human resources specialists, particularly recruitment specialists, travel to attend job fairs, visit college campuses, and meet with applicants.

Most specialists work full time during regular business hours. Some work more than 40 hours per week. Human resources specialists who have a thorough knowledge of their organization and its personnel regulations may advance to become human resources managers. Specialists may increase their chance of advancement by taking on new responsibilities or completing voluntary certification programs.

## EMPLOYMENT AND OUTLOOK

Human resource specialists held about 782,800 in 2021. Employment of human resources specialists is projected to grow 8 percent from 2021 to 2031, faster than the average for all occupations.

About 81,900 openings for human resources specialists are projected each year, on average, over the decade. Many of those openings are expected to result from the need to replace workers who transfer to different occupations or exit the labor force, such as to retire.

Companies are likely to continue to outsource human resources functions to organizations that provide these services, rather than directly employing human resources specialists. In addition, the services of human resources generalists will likely be needed to handle increasingly complex employment laws and benefit options.

### Related Occupations

- Compensation/Benefits Manager
- Compensation/Benefits/Job Analysist Specialist
- Customer Service Representative
- Human Resource Manager
- Insurance Sales Agent
- Labor Relations Specialist
- Public Relations Specialist
- Tax Examiner/Collector/Revenue Agent
- Training/Development Manager
- Training/Development Specialist

## MORE INFORMATION

**HR Certification Institute (HRCI)**
1725 Duke Street, Suite 400
Alexandria, VA 22314
866.898.4724
info@hrci.org
www.hrci.org

**International Foundation of Employee Benefit Plans (IFEBP)**
18700 W. Bluemound Road
Brookfield, WI 53045
888.334.3327
www.ifebp.org

**International Public Management Association for Human Resources (IPMA-HR)**
1617 Duke Street
Alexandria, VA 22314
703.549.7100
www.ipma-hr.org

**Society for Human Resource Management (SHRM)**
1800 Duke Street
Alexandria, VA 22314
800.283.7476
shrm@shrm.org
www.shrm.org

**World at Work**
14040 N. Northsight Boulevard
Scottsdale, AZ 85260
877.951.9191
customerexperience@worldatwork.org
www.worldatwork.org

---

*Daniel Vest*

# Information Security Analyst

## Snapshot

**Career Cluster(s):** Business Management & Administration; Information Technology

**Interests:** Computers; Data; Statistics; Security

**Earnings (Yearly Average):** $102,600

**Employment & Outlook:** Much Faster Than Average Growth Expected

## OVERVIEW

### Sphere of Work

Information security analysts plan and carry out security measures to protect an organization's computer networks and systems. As the number of organizations that do business online increases, so do the chances of cyberthreats increase. It is up to these analysts to create innovative solutions to prevent hackers from stealing critical information or creating problems for computer networks.

### Work Environment

Most information security analysts work for computer companies, consulting firms, or business and financial companies.

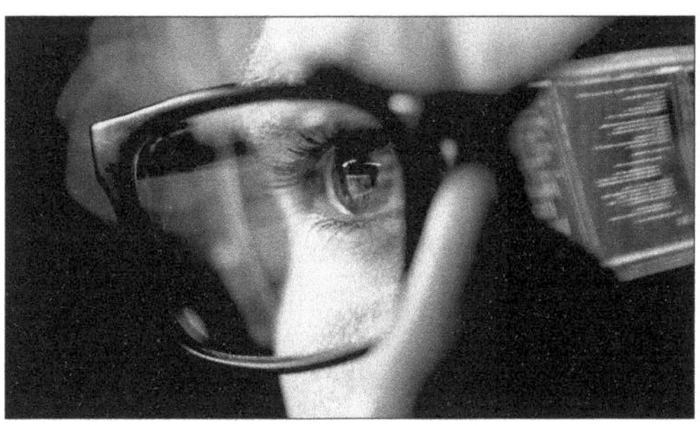

Photo via iStock.com/gorodenkoff. [Used under license.]

### Occupation Interest

Information security analysts are by nature conventional, investigative, and realistic-minded. They prefer working in a profession where they can follow a set of data to logical conclusions. Cybersecurity is a primary interest, as increasingly so many companies do their business online. These professionals will have to innovate solutions to the problems posed by potential malicious hackers, ransomware, and other cyberattacks.

### Profile

**Working Conditions:** Inside
**Education Needs:** Bachelor's Degree
**Licensure/Certification:** Security+; CISSP
**Opportunities for Experience:** Relevant industry experience or related jobs
**Interest Score:** CIR

### A Day in the Life—Duties and Responsibilities

Information security analysts plan and carry out security measures to protect an organization's computer networks and systems. They are heavily involved with creating their organization's disaster recovery plan, a procedure that IT employees follow in case of emergency. These plans allow for the continued operation of an organization's IT department. The recovery plan includes preventive measures such as regularly copying and transferring data to an offsite location. It also involves plans to restore proper IT functioning after a disaster. Analysts continually test the steps in their recovery plans.

Information security analysts must stay up to date on IT security and on the latest methods attackers are using to infiltrate computer systems. Analysts need to research new security technology to decide what will most effectively protect their organization.

### Duties and Responsibilities

- Monitoring their organization's networks for security breaches and investigate when one occurs
- Using and maintaining software, such as firewalls and data encryption programs, to protect sensitive information
- Checking for vulnerabilities in computer and network systems
- Researching the latest information technology (IT) security trends
- Preparing reports that document general metrics, attempted attacks, and security breaches
- Developing security standards and best practices for their organization
- Recommending security enhancements to management or senior IT staff
- Helping computer users when they need to install or learn about new security products and procedures

## WORK ENVIRONMENT

### Immediate Physical Environment

Information security analysts will work in office environments, with many hours logged in computer/desk work. Most information security analysts work full time, and some work more than 40 hours per week. Information security analysts sometimes have to be on call outside of normal business hours in case of an emergency.

### Human Environment

Information security analysts work alongside others and sometimes perform in leadership roles. They will find themselves having to explain highly technical information in non-technical terms for the benefit of others in their organization. A facility with words and strong communication skills are required.

### Technological Environment

This profession requires an in-depth knowledge of various computer systems and networks, programming languages and software. Security programs these analysts deal in and innovate will be the latest, cutting-edge programs possible, in order to combat the ever-evolving nature of the cybercrime that threatens organizations.

## EDUCATION AND TRAINING

### High School/Secondary

High school students interested in becoming information security analysts should focus on math, statistics, and computers-related courses. Becoming an information security analyst will probably requires a bachelor's degree, so students should prepare themselves for postsecondary education, including researching the schools they would like to consider attending to achieve that goal.

**Suggested High School Subjects**

- Algebra
- Computer Science
- English
- Pre-Calculus
- Psychology
- Statistics

**Related Career Pathways/Majors**

*Business Management & Administration Career Cluster*

- Business Information Management Pathway

*Information Technology Career Cluster*

- Network Systems Pathway

- Information Support & Services Pathway

*Law, Public Safety, Corrections & Security Career Cluster*

- Security & Protective Services Pathway

### Postsecondary

Information security analysts typically need a bachelor's degree in computer and information technology or a related field, such as engineering or math. However, some workers enter the occupation with a high school diploma and relevant industry training and certifications.

**Related College Majors**

- Business Management & Administration
- Computer Science
- General Business
- Information Technology
- International Business
- Mathematics

### *Transferable Skills and Abilities*

**Analytical Skills**

- Studying computer systems and networks
- Assessing risks to determine improvements for security policies and protocols

**Communication Skills**

- Explaining information security needs and potential threats to technical and nontechnical audiences within their organizations

**Creative Skills**

- Anticipating information security risks
- Implementing new ways to protect their organizations' computer systems and networks

**Detail-oriented**

- Paying careful attention to computer systems and watching for minor changes in performance in advance of otherwise hard-to-detect cyberattacks

**Problem-solving Skills**

- Responding to security alerts and uncovering and fixing flaws in computer systems and networks

### Adult Job Seekers

Information security analysts may need to have work experience in a related occupation. Many analysts have experience in an information technology department, often as a network and computer systems administrator.

### Professional Certification and Licensure

Many employers prefer to hire candidates who have information security certification. Some of these certifications, such as Security+, are for workers at the entry level; others, such as the Certified Information Systems Security Professional (CISSP), are designed for experienced information security workers. Certification in specialized areas, such as systems auditing, also is available.

## EARNINGS AND ADVANCEMENT

The median annual wage for information security analysts was $102,600 in 2021. The lowest 10 percent earned less than $61,520, and the highest 10 percent earned more than $165,920.

Information security analysts may advance to become chief security officers or another type of computer and information systems manager. Information security analysts also may advance within the occupation as they gain experience. For example, they may lead a team of other information security analysts or become an expert in a particular area of information security.

## EMPLOYMENT AND OUTLOOK

Information security analysts held about 163,000 jobs in 2021. Employment is projected to grow 35 percent from 2021 to 2031, much faster than the average for all occupations.

About 19,500 openings for information security analysts are projected each year, on average, over the decade. Many of those openings are expected to result from the need to replace workers who transfer to different occupations or exit the labor force, such as to retire.

High demand is expected for information security analysts. Cyberattacks have grown in frequency, and these analysts will be needed to create innovative solutions to prevent hackers from stealing critical information or creating problems for computer networks.

As businesses focus on enhancing cybersecurity, they will need information security analysts to secure new technologies from outside threats or hacks. A shift to remote work and the rise of e-commerce have increased the need for enhanced security, contributing to the projected employment growth of these workers over the decade.

Strong growth in digital health services and telehealth will also increase data security risks for healthcare providers. More of these analysts are likely to be needed to safeguard patients' personal information and data.

## Fast Fact

For those starting out their careers as cybersecurity analysts, one way to gain some experience is to attend as many hackathons as they can. These events will give them a front-row seat to see how the best of the best in cybersecurity approach their work. And, they make for excellent networking opportunities that may come in handy later. Source: thehackernews.com

Photo via iStock.com/ra2studio. [Used under license.]

### Related Occupations

- Computer/Information Research Scientist
- Computer/Information Systems Manager
- Computer Network Architect
- Computer Programmer
- Computer Support Specialist
- Computer Systems Analyst
- Database Administrator/Architect
- Network/Computer Systems Administrator
- Software Developer/Quality Assurance Analyst/Tester
- Web Developer/Digital Designer

## MORE INFORMATION

**Association for Computing Machinery (ACM)**
1601 Broadway, 10th Floor
New York, NY 10019-7434
212.869.7440
acmhelp@acm.org
www.acm.org

**Association for Women in Computing (AWC)**
P.O. Box 2768
Oakland, CA 94602
info@awc-hq.org
www.awc-hq.org

**Computing Research Association (CRA)**
1828 L Street NW, Suite 800
Washington, DC 20036
202.234.2111
cra.org

**IEEE Computer Society**
2001 L Street NW, Suite 700
Washington, DC 20036
202.371.0101
help@computer.orrg
computer.org

**International Association of Computer Science and Information Technology (IACSIT)**
contact@iacsit.org
www.iacsit.org

**National Center for Women & Information Technology (NCWIT)**
1909 26th Street, 2nd Floor
Boulder, CO 80302
303.735.6671
info@ncwit.org
ncwit.org

**USA JOBS**
1900 E Street NW
Washington, DC 20415
www.usajobs.gov

---

*Daniel Vest*

## *Conversation With...*

# RYDER JEFFERSON MOSES

Vice President of Information Security
Electronic Payment Processor, Mid-Atlantic Region
IT, 32 years; Cybersecurity, 5 years

**What was your individual career path in terms of education/training, entry-level job, or other significant opportunity?**

I graduated with a biology degree in 1981 to limited employment opportunities, so I fixed computers and printers because I have good mechanical aptitude. I realized that what really grabbed my attention was networks, data transmission, and telecommunications carrier services between computers. This was before the internet. As my career in network engineering progressed, I experienced, first-hand, the evolution of data transmission protocols for both private and public (internet) networks.

I wound up working as a contractor on government networks, and decided to go back to school so I had the academic credentials to back up the work I'd been doing for 10 years. I got my master's in Telecommunications Management; today I would recommend a degree in information security, which wasn't available at that time.

I went on to work for a company that supports banks in the electronic payment processing industry. As that business grew—dramatically—and credit card processing migrated toward the internet, it became vulnerable to hackers who stole information to produce counterfeit credit cards. This was in the early 2000s, and the increased use of the internet gave rise to new encryption methods for data transmission. As I spent more time implementing encryption methods, I was working closely with people in the developing field of information security.

As encryption grew more sophisticated, the hackers became more sophisticated at uncovering vulnerabilities, which led to the creation of a security standard to force banks, merchants, and processors to improve data protection methods. In my most recent role, I helped explain to merchants the need to implement these measures to meet the Payment Card Industry Data Security Standards. I also helped merchants make changes to support chip-enabled payment cards.

**What are the most important skills and/or qualities for someone in your profession?**

Critical thinking skills, an analytical approach to risk assessment, and a fundamental understanding of cryptography and data transmissions protocols, because you need the ability to do an end-to-end assessment of the data processing environment. Also, you need a clean background check with no arrests and an absolute dedication to integrity. You aren't going to get a join information security if potential employers have any reason to suspect they can't trust you.

**What do you wish you had known going into this profession?**

The fact that you will be on call 24/7, 365 days because you need to drop whatever you're doing and devote all your attention to any kind of data compromise. I once left a company meeting to help decode transaction data in real time after a guy who'd just gotten out of jail for committing credit card fraud stole a dial-up terminal and started doing fraudulent credit card transactions from his house. I was on a conference call with law enforcement sitting in a van across from his house as I manually decoded his fraudulent transactions so they could catch him in the act.

**Are there many job opportunities in your profession? In what specific areas?**

Definitely. Every merchant who accepts credit and debit cards needs technical personnel to assess vulnerabilities and implement measures to protect themselves, the cardholders, and the banks. Many industries face growing information security requirements, including banks, electronic funds processing companies, stock brokerages, hospitals, government agencies, and colleges.

**How do you see your profession changing in the next five years, and what skills will be required?**

In this business, the technology—and the use of technology by the bad guys—evolves at a rapid pace. You have to constantly update your knowledge and skills, and continuously monitor systems and networks for emerging trends in vulnerability exploitation.

**What do you enjoy most about your job? What do you enjoy least about your job?**

I enjoy the satisfaction of knowing I'm helping to protect financial data and reduce the risk of financial losses to cardholders, merchants, and banks. I also enjoy explaining very technical security requirements in terms that everyone can understand. This field pays well, and I enjoy that.

I don't like being on call. I don't like security breaches, and the stress levels associated with protecting the country's commerce infrastructure while knowing that information security is only effective until the bad guys find another vulnerability to exploit.

**Can you suggest a valuable "try this" for students considering a career in your profession?**

Internships are a great way to learn about networks and data transmission technologies. There's also, certainly, a lot of information online. Look into things like Cisco router training. Also, the military is worth considering because they have training opportunities due to the critical requirements for keeping data secure.

*This interview was originally published in 2016.*

# Insurance Underwriter

**Snapshot**

**Career Cluster(s):** Consumer Services; Finance

**Interests:** Accounting; Finance; Data

**Earnings (Yearly Average):** $76,390

**Employment & Outlook:** Decline In Growth Expected

## OVERVIEW

### Sphere of Work

Insurance underwriters decide whether to provide insurance, and under what terms. They evaluate insurance applications and determine coverage amounts and premiums. In an increasingly online world, insurance firms use the process of "cyber insurance underwriting" to determine their clients' risk, their exposure, and how much it will cost to recover from a data breach, a ransomware attack, or other harmful cyber activity.

### Work Environment

Underwriters work indoors in offices. Although underwriters spend most of their time working alone on applications at a computer, they sometimes must handle customer inquiries.

Some property and casualty underwriters may travel to assess properties in person. Most underwriters work full time.

Photo via iStock.com/blackred. [Used under license.]

### Occupation Interest

Insurance underwriters enjoy working with numbers, taking and analyzing data to arrive at conclusions. They are methodical, practical and thorough in approach. Conventional by nature, they can enjoy long stretches of solitary work and are capable of making difficult decisions. In the case of cyber insurance underwriting, these professionals have a knack for the latest technology and software systems.

### Profile

**Working Conditions:** Inside
**Education Needs:** Bachelor's Degree
**Licensure/Certification:** Various
**Opportunities for Experience:** Coursework; related work experience
**Interest Score:** CEI

## A Day in the Life—Duties and Responsibilities

Underwriters are the main link between an insurance company and an insurance agent. Insurance underwriters use computer software programs to determine whether to approve an applicant. They take specific information about a client and enter it into a program. The program then provides recommendations on coverage and premiums. Underwriters evaluate these recommendations and decide whether to approve or reject the application. If a decision is difficult, they may consult additional sources, such as medical documents and credit scores.

For simple and common types of insurance, such as automobile insurance, underwriters can typically rely on automated recommendations. For more specific and complex insurance types, such as workers' compensation, underwriters need to rely more on their own analytical insight.

Underwriters analyze the risk factors appearing on an application. For instance, if an applicant reports a previous bankruptcy, the underwriter must determine whether that information is relevant to the policy being applied for. The underwriter would likely consider how far in the past the bankruptcy occurred and how the applicant's financial situation has changed since the applicant filed for bankruptcy.

Insurance underwriters must achieve a balance between risky and cautious decisions. If underwriters allow too much risk, the insurance company will pay out too many claims. But if they don't approve enough applications, the company will not make enough money from premiums.

### Duties and Responsibilities

- Analyzing information stated on insurance applications
- Determining the risk involved in insuring a client
- Screening applicants on the basis of set criteria
- Evaluating recommendations from underwriting software
- Contacting field representatives, medical personnel, and others to obtain further information
- Deciding whether to offer insurance
- Determining appropriate premiums and amounts of coverage
- Reviewing and updating the rules that govern automation software

Most insurance underwriters specialize in one of three broad fields: life, health, and property and casualty. Although the job duties in each field are similar, the criteria that underwriters use vary. For example, for someone seeking life insurance, underwriters consider the person's age and financial history. For someone applying for car insurance (a form of property and casualty insurance), underwriters consider the person's driving record.

Within the broad field of property and casualty, underwriters may specialize even further into commercial (business) insurance or personal insurance. They may also specialize by the type of policy, such as for automobiles, boats (marine insurance), or homes (homeowners' insurance).

## WORK ENVIRONMENT

### Immediate Physical Environment

Underwriters spend most of their time in an office working behind a computer at their desk. They'll usually take several hours to input applicant data or complete calculations to assess risks. Some underwriters will travel to the location of an agent or client to discuss more complex work items or negotiate contract terms in person.

### Human Environment

When they're not completing their more clerical tasks, insurance underwriters will typically work in collaboration with other employees, clients or insurance agents via phone calls or emails. Strong communication and interpersonal skills are therefore a must for the job.

### Technological Environment

Insurance underwriters use computer software programs to determine whether an applicant should be approved. The variety of software they may use in the daily course of their work ranges from financial analysis software, to presentation software, and includes many more types still.

## EDUCATION AND TRAINING

### High School/Secondary

High school students interested in becoming insurance underwriters should focus on math, statistics, and business-related courses. Becoming an underwriter requires a bachelor's degree, so students should prepare themselves for postsecondary education, including researching the schools they would like to consider attending to achieve that goal.

**Suggested High School Subjects**

- Algebra
- Civics
- Economics
- Pre-Calculus
- Statistics

**Related Career Pathways/Majors**

*Finance Career Cluster*

- Insurance Pathway

*Human Services Career Cluster*

- Consumer Services Pathway

### Postsecondary

Insurance underwriters typically need a bachelor's degree to enter the occupation. However, insurance-related work experience and strong computer skills may be enough for some positions. Certification is generally necessary for advancement to senior underwriter and underwriter manager positions.

Employers usually prefer to hire candidates who have a bachelor's degree. A common field of degree is business. Coursework in finance, economics, and mathematics is helpful.

Beginning underwriters usually work as trainees under the supervision of senior underwriters. Trainees work on basic applications and learn the most common risk factors. Some companies offer training programs that include classroom instruction on the basics of underwriting.

#### *Transferable Skills and Abilities*

**Analytical Skills**

- Evaluating information from a variety of sources
- Solving complex problems.

**Decision-making Skills**

- Deciding whether to offer insurance coverage
- Deciding at what level to set premiums

**Detail-oriented**

- Paying attention to detail, because each individual item on an insurance application can affect the coverage decision

**Interpersonal Skills**

- Having good communication and interpersonal skills because much of their work involves dealing with other people, such as insurance agents.

**Math Skills**

- Determining the probability of losses on an insurance policy and calculating appropriate premiums require mathematical ability

## Fast Fact

**The insurance industry is a prime target for cyberattacks. A recent industry report states that 20% of the top 99 insurance carriers have a high rate of vulnerability to ransomware, and the rate of hacks have increased by 300% in the last year alone. The largest ransom paid to date by an insurance company totaled $40 million, with the average**

Photo via iStock.com/manopjk. [Used under license.]

As new underwriters gain experience, they work independently and handle more complex applications.

**Related College Majors**

- Accounting
- Auditing
- Business Management & Administration
- Economics
- Entrepreneurial Studies
- Financial Management
- General Business
- Insurance
- International Business

### Adult Job Seekers

Adults seeking to become insurance underwriters should have a background or strength in business or finance. Schooling is still required, so thought should be given to an individual's long-term professional goals.

### Professional Certification and Licensure

Employers often expect underwriters to become certified through coursework. These courses are important for keeping current with new insurance policies and for adjusting to new technology and changes in state and federal regulations. Certification is often necessary for advancement to senior underwriter and underwriter management positions. Many certification options are available.

For underwriters with at least 2 years of insurance experience, The Institutes offer the Chartered Property and Casualty Underwriter (CPCU) designation. For beginning underwriters, The Institutes offer a training program.

The Institutes also offer several other designations in insurance specialties, including the Associate in Commercial Underwriting (AU) and Associate in Personal Insurance (API). To earn these designations, underwriters complete a series of courses and exams that generally takes 1 to 2 years.

The National Association of Insurance and Financial Advisors offers the Life Underwriter Training Council Fellow (LUTCF) designation, which consists of a three-part curriculum in basic insurance concepts.

The American College of Financial Services offers the Chartered Life Underwriter (CLU) certification. This certification consists of five core courses and three electives, and candidates must have 3 years of related work experience.

## EARNINGS AND ADVANCEMENT

The median annual wage for insurance underwriters was $76,390 in 2021. The lowest 10 percent earned less than $47,330, and the highest 10 percent earned more than $126,380.

## EMPLOYMENT AND OUTLOOK

Insurance underwriters held about 123,300 jobs in 2021. Employment of insurance underwriters is projected to decline 4 percent from 2021 to 2031.

Despite declining employment, about 8,400 openings for insurance underwriters are projected each year, on average, over the decade. All of those openings are expected to result from the need to replace workers who transfer to other occupations or exit the labor force, such as to retire.

Automated underwriting software allows workers to process applications quickly, reducing the need for underwriters. As this technology continues to improve and become more widely adopted in the insurance industry, more underwriting decisions are expected to be made automatically.

However, there still will be a need for underwriters to review and update the criteria that run the automation. In addition, their analytical insight will be needed in specific fields, such as workers' compensation, marine insurance, and health insurance.

### Related Occupations

- Actuary
- Budget Analyst
- Claims Adjuster/Appraiser/Examiner/Investigator
- Cost Estimator
- Insurance Sales Agent
- Loan Officer

## MORE INFORMATION

**The American College of Financial Services**
630 Allendale Road, Suite 400
King of Prussia, PA 19406
610.526.1000
866.216.8112
www.theamericancollege.edu

**Insurance Information Institute (III)**
110 William Street
New York, NY 10038
212.346.5500
info@iii.org
www.iii.org

**The Institutes**
720 Providence Road
Malvern, PA 19355
800.644.2101
customersuccess@theinstitutes.org
web.theinstitutes.org

**National Association of Insurance and Financial Advisors (NAIFA)**
2901 Telestar Court
Falls Church, VA 22042
877.866.2432
info@naifa.org
belong.naifa.org

*Daniel Vest*

## *Conversation With...*
## MICHAEL KOSCIELNY

Director of Underwriting, Florida Peninsula Insurance
Boca Raton, Florida
Insurance underwriting, 35 years

**What was your individual career path in terms of education/training, entry-level job, or other significant opportunity?**

After graduating from college with a degree in economic history, my dream was to work for the Federal Reserve. After a short stint as a bank investment officer, I found my way to the insurance industry as a salesperson for Metropolitan Life. Working late evenings and most Saturdays was not my idea of an enticing career path. I was fairly successful, but I needed a job with regular hours to keep my new wife happy. Through a recruiter who specialized in insurance, I found myself interviewing at SAFECO Insurance Company. I had no idea who they were, but the hours were 8 to 5, Monday through Friday. The job was a personal lines underwriter. After one of the longest interview processes I have ever experienced, I got the job. Little did I know at the time what landing at SAFECO would mean to me. It was a company known for training at every level of the organization, and underwriting was the place to be. Underwriters analyze information to determine the risk associated with insuring a potential client. Using guidelines established by our company, we examine the characteristics of the risk presented to ensure that the coverage and the premium charged is appropriate. The goal is to predict with some certainty the potential for loss within the tolerances of the company.

In those early years, I learned the skills necessary to be a successful underwriter. Not only did SAFECO have top-notch training, but they also supported education through the AICPCU, now known as The Institutes. My "learning curve" was actually a vertical line for many years, allowing me to earn the Chartered Property Casualty Underwriter (CPCU) designation, Certified Insurance Counselor (CIC), and many others. In my twenty years with SAFECO, I never held the same job more than five years, progressing up the management ranks and crossing over to marketing and sales and back to underwriting. I am currently president-elect of the CPCU Society Leadership Council.

**What are the most important skills and/or qualities for someone in your profession?**

I would sum up the skills necessary to be a great underwriter with two words: intellectual curiosity. Of course you have to have good communication skills, time-management skills, analytical skills, decisiveness, good judgment and be a team player, but intellectual curiosity is the key. You have to be nosey and question things that just don't make sense. If you love to solve problems, underwriting is the place to be.

**What do you wish you had known going into this profession?**
I wish I had understood while I was in college that insurance includes many careers beyond sales. I wouldn't have wasted two years of my life in the banking industry if someone had shared with me the opportunities in insurance.

**Are there many job opportunities in your profession? In what specific areas?**
The opportunities over the next ten years are huge. One third of the 3 million people working in the insurance industry will retire in the next decade. That means great opportunity for young people to find their way into this industry and succeed. There will be opportunities in all facets of insurance company operations, not just underwriting.

**How do you see your profession changing in the next five years? What role will technology play in those changes, and what skills will be required?**
Technology has removed the mundane from underwriters' work, allowing them to focus on more complex decisions. Routine risk assessments are done by computers, providing potential clients with faster decisions. Risks that cannot be processed electronically are steered to an underwriter. It's a very rewarding job today.

**What do you enjoy most about your job? What do you enjoy least about your job?**
Underwriting—or the systematic evaluation of risks and how I might mitigate, accept or avoid the consequences of those risks—has become a way of life for me. It's a routine I use in every facet of what I do, professionally and personally. I love that every risk I evaluate is a new story with unique twists. I can honestly say there is nothing I dislike about my job. I guess what I like least are those things that do not challenge me. With any job, there are routine things that make it a "job," like paperwork. Not interesting, but necessary.

**Can you suggest a valuable "try this" for students considering a career in your profession?**
Get involved in a degree program at a school with risk management courses. There are many across the country.

Check out insuremypath.org. You can get connected with internships at top insurance companies. There's nothing like taking a career for a test drive before you make the leap. If you're researching colleges, look for one with risk management courses. Finally, reach out to the CPCU Society. They have more than 22,000 members across the country who are more than willing to spend time with students interested in this great industry.

*This interview was originally published in 2016.*

# Loan Officer

**Snapshot**

**Career Cluster(s):** Business, Management & Administration; Finance

**Interests:** Business; Finance

**Earnings (Yearly Average):** $63,380

**Employment & Outlook:** As Fast As Average Growth Expected

## OVERVIEW

### Sphere of Work

Loan officers evaluate, authorize, or recommend approval of loan applications for people and businesses using a process called underwriting to assess whether applicants qualify for loans. Consumer loan officers specialize in loans to people, whereas commercial loan officers specialize in loans to businesses. Business loans are often essential to the instigation of, or continuance of, a business' operations. Further, mortgage loan officers specialize in loans used to buy real estate. Because of the sensitive financial information involved, cybersecurity risks like loss of client information or system access issues are always to be considered. The loan officer themselves or someone on their team must be appointed responsible for overseeing cybersecurity risk to ensure accountability.

### Work Environment

Loan officers who specialize in consumer loans usually work in offices. Mortgage and commercial loan officers often work outside the office and meet with clients at their homes or businesses, as a means of assessing the client and viability of the property or business.

Photo via iStock.com/skynesher. [Used under license.]

## Occupation Interest

Loan officers have an interest in finance, math, and statistics, as well as a desire to help members of the public achieve financial and personal goals, while ensuring that the company for which the loan officer works is well-served. Loan officers make decisions that greatly affect peoples' lives—be it when securing a first mortgage, or a startup loan for an entrepreneurial endeavor.

### Profile

**Working Conditions:** Inside
**Education Needs:** On-the-Job Training; Bachelor's Degree
**Licensure/Certification:** Required; Varies
**Opportunities for Experience:** Internship
**Interest Score:** CES

## A Day in the Life—Duties and Responsibilities

Loan officers use a process called underwriting to assess whether applicants qualify for loans. After collecting and verifying all the required financial documents, the loan officer evaluates the information they obtain to determine the applicant's need for a loan and ability to pay back the loan. Most firms use underwriting software, which produces a recommendation for the loan based on the applicant's financial status. After the underwriting software produces a recommendation, loan officers review the output of the software and consider any additional information to make a final decision.

### Duties and Responsibilities

- Contacting companies or people to ask if they need a loan
- Meeting with loan applicants to gather personal information and answer questions
- Explaining different types of loans and the terms of each type to applicants
- Obtaining, verifying, and analyzing the applicant's financial information, such as credit rating and income level
- Reviewing loan agreements to ensure that they comply with federal and state regulations
- Approving loan applications or referring them to management for a decision

The work of loan officers has sizable customer-service and sales components. Loan officers often answer questions and guide customers through the application process. In addition, many loan officers must market the products and services of their lending institution and actively solicit new business.

Consumer loan officers specialize in loans to people. Consumers take out loans for many reasons, such as buying a car or paying college tuition. For some simple consumer loans, the underwriting process is fully automated. However, the loan officer is still needed to guide applicants through the process and to deal with cases with unusual circumstances. Some institutions—generally small banks and credit unions—do not use underwriting software and instead rely on loan officers to complete the underwriting process manually.

Commercial loan officers specialize in loans to businesses, which often use the loans to begin operations, buy supplies, and upgrade or expand existing operations. Commercial loans often are larger and more complicated than other types of loans. Because companies have such complex financial situations and statements, commercial loans typically require human judgment in addition to the analysis by underwriting software. Furthermore, some commercial loans are so large that no single bank will provide the entire amount requested. In such cases, loan officers may have to work with multiple banks to put together a package of loans.

Mortgage loan officers specialize in loans used to buy real estate (property and buildings), which are called mortgage loans. Mortgage loan officers work on loans for both residential and commercial properties. Often, mortgage loan officers must seek out clients, which requires developing relationships with real estate companies and other sources that can refer prospective applicants.

## OCCUPATION SPECIALTIES

### Loan Collection Officer

Loan collection officers contact borrowers who fail to make their loan payments on time. They work with borrowers to help them find a way to keep paying off the loan. If the borrower continues to miss payments, loan officers start the process of taking away what the borrower used to secure the loan (called "collateral")—often a home or car—and selling it to repay the loan.

### Loan Underwriter

Loan underwriters specialize in evaluating whether a client is creditworthy. They collect, verify, and evaluate the client's financial information provided on their loan application and then use loan underwriting software to produce recommendations.

## WORK ENVIRONMENT

### Immediate Physical Environment

Loan officers working on the consumer side spend most of their time in offices. Commercial loan officers, however, may spend quite a bit of time out of the office, traveling to meet applicants and assess properties and businesses in person. Most loan officers work full-time and some work extensive hours.

### Human Environment

While some loan officers may spend most of their time in an office working with software to generate recommendations, others require a person-to-person approach in order to assess and prepare the loan. Loan officers must exhibit good judgment and business sense, while also understanding that the decisions they make directly impact the lives of their clients. They are the gateway to funds that will help their clients to grow and succeed in life and in business, but the loans must also be profitable. In many cases loan officers need to "sell" themselves to attract business, whether by directly interacting with prospective clients, or networking with real estate agents or other firms.

### Technological Environment

Computer and software usage is common, including specialized underwriting software, as well as database software, accounting software, and platforms that allow the loan officer to access credit ratings. The officer must be extremely familiar with these programs to carry out day-to-day functions without error.

## EDUCATION AND TRAINING

### High School/Secondary

High school students with an interest in entering the world of banking and finance should focus on math, economics, and statistics courses, while ensuring good grades in English to improve communication skills. Students should think ahead to postsecondary, in terms of what pathways and schools will help meet their career goals.

**Suggested High School Subjects**

- Algebra
- Biology
- Calculus
- Chemistry
- Earth or Life or Physical Science
- Economics
- English
- Geometry
- History
- Physics
- Psychology

- Statistics
- Trigonometry

**Related Career Pathways/Majors**

*Business Management & Administration Career Cluster*

- Business Information Management Pathway

*Finance Career Cluster*

- Accounting Pathway
- Banking Services Pathway
- Business Finance Pathway

## Postsecondary

Loan officers typically need a bachelor's degree, usually in a field such as business or finance. Because commercial loan officers analyze the finances of businesses applying for credit, they need to understand general business accounting, including how to read financial statements.

**Related College Majors**

- Business
- Economics
- Finance

## Adult Job Seekers

Some jobseekers may be able to enter the occupation without a bachelor's degree if they have related work experience, such as experience in sales, customer service, or banking.

## Professional Certification and Licensure

Mortgage loan officers must have a Mortgage Loan Originator (MLO) license. To become licensed, they must complete at least 20 hours of coursework, pass an exam, and submit to background and credit checks. Licenses must be renewed annually, and individual states may have additional requirements.

Several banking associations, including the American Bankers Association (ABA) and the Mortgage Bankers Association (MBA), as well as a number of schools, offer courses, training programs, or certifications for loan officers. Al-

### *Transferable Skills and Abilities*

**Decision-making Skills**

- Assessing an applicant's financial information and deciding whether to award the applicant a loan

**Detail-oriented**

- Paying attention to detail, as each piece of information on an application can have a major effect on the profitability of a loan

**Initiative**

- Seeking out new clients
- Promoting their lending institution and contacting people and firms to determine their need for a loan

**Interpersonal Skills**

- Guiding customers through the application process and answer their questions though not required, certification shows dedication and expertise and, thus, may enhance a candidate's employment opportunities.

### Additional Requirements

Once hired, loan officers usually receive some on-the-job training. This may be a combination of formal, company-sponsored training and informal training during the first few months on the job.

## EARNINGS AND ADVANCEMENT

Earnings depend on the experience of the loan officer, the size of the firm for which they work, and their area of specialty. Median annual earnings of loan officers were $63,380 in 2021. The lowest 10 percent earned less than $32,520, and the highest 10 percent earned more than $138,310.

The form of compensation varies widely by employer. Some loan officers are paid a flat salary; others are paid on commission. Those on commission usually are paid a base salary plus a commission for the loans they originate. Loan officers also may receive extra commission or bonuses based on the number of loans they originate or how well the loans perform.

## EMPLOYMENT AND OUTLOOK

Loan officers held 354,600 jobs in 2021. Employment is expected to grow by 4 percent through the year 2031, about as fast as the average for all occupations. Increased demand for loan officers is expected as both businesses and individuals seek credit to finance commercial investments and personal spending. Loan officers will be needed to evaluate the creditworthiness of applicants and to determine the likelihood that loans will be paid back in full and on time.

However, the decline of bank branches and the increased use of productivity-enhancing technology in loan processing are expected to slow employment growth.

Job opportunities should be good for people with lending, banking, or sales experience. In addition, some firms require loan officers to find their own clients, so candidates with established contacts and a referral network should have the best job opportunities.

## Fast Fact

**The work-from-home lifestyle brought on by COVID-19 had serious ramifications for the financial world, as it inspired more hacking than ever before, with a 238% rate of increase in cyberattacks against banks between February and April 2020. Banking professionals including loan officers should be aware of the ongoing potential dangers, as experts claim U.S. banks are particularly vulnerable to state-sponsored cyberattacks from nations like North Korea, China, and Russia.** Source: investopedia.com

Photo via iStock.com/andresr. [Used under license.]

### Related Occupations

- Financial Analyst
- Financial Examiner
- Financial Manager
- Insurance Sales Agent
- Personal Financial Advisor
- Real Estate Broker/Sales Agent
- Securities/Commodities/Financial Services Sales Agent
- Tax Examiners/Collectors/Revenue Agent
- Teller

## MORE INFORMATION

**American Bankers Association (ABA)**
1120 Connecticut Avenue NW
Washington, DC 20036
800.226.5377
www.aba.com

**Mortgage Bankers Association (MBA)**
1919 M Street NW, 5th floor
Washington, DC 20036
202.557.2700
membership@mba.org
www.mba.org

**National Association of Mortgage Brokers (NAMB)**
601 Pennsylvania Avenue NW, South Building
Washington, DC 20004
202.434.8250
membership@namb.org
namb.org

**National Association of Mortgage Processors (NAMP)**
www.mortgageprocessor.org

**National Association of Mortgage Underwriters (NAMU)**
Presidential Circle
4000 Hollywood Boulevard, Suite 555-South
Hollywood, FL 33021
800.865.8550
contact@mortgage-underwriters.org
www.Mortgage-Underwriters.org

**Nationwide Multistate Licensing System & Registry (NMLS) Resource Center**
855.665.7123
mortgage.nationwidelicensingsystem.org

---

*Stuart Paterson; updated by Daniel Vest*

# *Conversation With...*

## PAUL WALSH

Mortgage Officer
Mortgage Financial, Inc.
Lending profession, 24 years

**What was your individual career path in terms of education/training, entry-level job, or other significant opportunity?**

I was actually a real estate broker for five years before going into lending. The experience of buying my first home made me decide to switch careers. (I actually, in a previous life, worked in high tech). When I bought my house, my financing experience was horrendous. After I closed on the house, I kept thinking about this for two or three years. Finally, I just made the change and left the world of high tech. So, it was both the personal experience I had and the opportunity I saw that led me to get into real estate and then lending. Also, I had my bachelor's degree and to move to the next level in the world of high tech would have required a master's degree. I had a family and it just wasn't in me to pursue a master's. There's really no training required of loan officers, which has to change, it just has to. But we are licensed and required to pass two exams.

**What are the most important skills and/or qualities for someone in your profession?**

Truly, right now it's understanding exactly what the consumer wants or needs and, to some extent, protecting them from themselves. If someone wants to buy a $450,000 house and just cannot afford it, part of the job is reeling them in with realistic expectations. You have to be able to place people in the right loan program. I may get compensated more if I put someone in an FHA loan versus a Fannie Mae, but is that the best thing for them? Or the loan with the lowest interest rate may not be the least expensive. So you just really have to know what you're doing—that sounds ridiculous, but you'd be amazed how many people in this field don't know what they're doing, even with a license. You have to have good ethics. You will be required to have a criminal background check, be fingerprinted by the FBI, and submit a credit report.

**What do you wish you had known going into this profession?**

I wish I had known the importance of having a team and working as a team. No one person can really do the entire loan process alone. For instance, I no longer do reverse mortgages. I now know my skill set. I wish I had known that I can't be all things to all people. Right from the start, I would have chosen a specialty and stuck to it.

**Are there many job opportunities in your profession? In what specific areas?**
Being a mortgage officer is really a sales job. You're selling a product, and those skills will always be in demand. But the future of this business is in the world of mortgage compliance, because the government is taking mortgage compliance very, very seriously. If sales isn't your thing, I would suggest compliance.

**How do you see your profession changing in the next five years, what role will technology play in those changes, and what skills will be required?**
Technology is huge. I'm very old school and like to meet person to person. I'm a pen and paper guy. But many mortgage officers or loan officers and their clients nowadays never meet. They rarely even talk. You just go online and submit everything electronically. e-Mortgages and e-financing is really where this business is going. I consider that unfortunate.

**What do you like most about your job? What do you like least about your job?**
For me, it's always the people, and first-time homebuyers in particular. What I enjoy most is the personal interaction, and basically solving problems for people. The economy hasn't been great. If someone has excessive credit card debt or other financial problems if they have a mortgage and a home—you can use that equity to structure things and really come up with a solution to help people out of unfortunate situations.

What I enjoy least is dealing with all the regulations, though I agree wholeheartedly with it. The compliance part is very, very difficult, and the inconsistencies from lender to lender and the multitude of product guidelines or "overlays" gets complicated.

**Can you suggest a valuable "try this" for students considering a career in your profession?**
Team up with someone who does what I do and work with them for a summer. You'll learn things like how to look at assets and recognize risk profiles. Even with an inside position, where you're sending out verification of employment forms, you'll learn the ropes and learn industry guidelines. Over the course of a summer, someone could learn a tremendous amount about this business.

*This interview was originally published in 2015.*

# Management Analyst

## Snapshot

**Career Cluster(s):** Business Management & Administration; Finance

**Interests:** Accounting; Business; Finance; Data; Statistics; Recordkeeping; Investigating

**Earnings (Yearly Average):** $93,000

**Employment & Outlook:** Much Faster Than Average Growth Expected

## OVERVIEW

### Sphere of Work

Management analysts are specialists who examine a company and suggest ways to boost productivity and profit. Before suggesting modifications to a company, they gather and examine data regarding its operations. Businesses use management analysts to examine various workflows and processes, increase productivity, create value, and experience a rise in revenue. In terms of cybersecurity, management analysts help support the oversight of first-line cyber defense efforts. They will monitor emerging cybersecurity threats and confer with senior management and other technology risk professionals, helping to refine existing cybersecurity practices and implement new strategies.

### Work Environment

Photo via iStock.com/NicolasMcComber. [Used under license.]

Management analysts usually divide their time between their offices and the client's site. Because they must spend a significant amount of time with clients, analysts travel frequently. Analysts may experience stress, especially when trying to meet a client's demands on a tight schedule. Analysts often work many hours under tight deadlines. Some work more than 40 hours per week.

### Profile

**Working Conditions:** Inside
**Education Needs:** Bachelor's Degree
**Licensure/Certification:** Various
**Opportunities for Experience:** Prior work experience in similar field
**Interest Score:** IEC

### Occupation Interest

Management analysts have analytical minds, with an ability to examine the facts and figures and come up with solutions to problems that bolster productivity and profit of companies they work for. Because it is also a very social job, where they are interacting with others often, strong communication skills are essential as they will need to relate complex ideas in clear and concise terms that are easily understandable.

### A Day in the Life—Duties and Responsibilities

Management analysts, often called management consultants, recommend ways to improve an organization's efficiency. They advise managers on how to make organizations more profitable through reduced costs and increased revenues.

Although some management analysts work for the organization that they analyze, many work as consultants on a contractual basis.

The work of management analysts may vary from project to project. Some projects require a team of analysts, each specializing in one area. On other projects, analysts work independently with the client organization's managers.

Management analysts often specialize in certain areas, such as inventory control or reorganizing corporate structures for efficiency. Some focus on a specific industry, such as healthcare or telecommunications. In government, management analysts usually specialize by type of agency.

Organizations hire management analysts to develop strategies for entering and remaining competitive in the market.

Management analysts who work on contract may write proposals and bid for jobs.

### Duties and Responsibilities

- Gathering and organizing information about the problems to be solved or the procedures to be improved
- Interviewing personnel and conducting onsite observations to determine the methods, equipment, and personnel that will be needed
- Analyzing financial and other data, including revenue, expenditure, and employment reports
- Developing solutions or alternative practices
- Recommending new systems, procedures, or organizational changes
- Making recommendations to management through presentations or written reports
- Conferring with managers to ensure changes are working

Typically, an organization that needs the help of a management analyst requests proposals from a number of consultants and consulting companies that specialize in the needed work. Interested companies then submit a proposal that explains details such as how the work will be completed, what the schedule will be, and how much it will cost. The organization selects the proposal that best meets its needs and budget.

## WORK ENVIRONMENT

### Immediate Physical Environment

Most management analysts work in offices, many of these in team environments. There may be travel involved, be it within the company itself or to various client locations in the course of their work.

### Human Environment

Management analysts will frequently work in teams, and need to have strong communication skills for their day-to-day activities and well as in instances where they're interacting with their management teams and/or clients.

### Technological Environment

Management analysts may use technology, such as data base user interface and query software systems, ERP software, among many other software types.

## EDUCATION AND TRAINING

### High School/Secondary

High school students interested in becoming management analysts should focus on math, and business-related courses. Becoming one requires at least a bachelor's degree, so students should prepare themselves for postsecondary education, including researching the schools they would like to consider attending to achieve that goal.

**Suggested High School Subjects**

- Civics
- Economics
- English
- Geometry

- Pre-Calculus
- Psychology
- Statistics

**Related Career Pathways/Majors**

*Business Management & Administration Career Cluster*

- General Management Pathway
- Business Information Management Pathway

*Government & Public Administration Career Cluster*

- Public Management & Administration Pathway

## Postsecondary

Management analysts typically need at least a bachelor's degree and several years of related work experience. A bachelor's degree is the typical entry-level requirement for management analysts. However, some employers prefer to hire candidates who have a master's degree in business administration (MBA).

### Transferable Skills and Abilities

**Analytical Skills**
- Interpreting information and using their findings to make proposals

**Communication Skills**
- Conveying information clearly in both writing and speaking
- Having good listening skills to understand an organization's problems
- Recommending appropriate solutions

**Interpersonal Skills**
- Working with managers and other employees of the organizations for which they provide consulting services
- Working as a team toward achieving the organization's goals

**Problem-solving Skills**
- Thinking creatively to solve clients' problems
- Solving the unique challenges associated with each new situation

**Time-management Skills**
- Working under tight deadlines
- Using their time efficiently to complete projects on schedule

Management analysts address a range of topics, and many fields of study provide a suitable educational background. Fields of bachelor's degree study may include business, social science, and engineering.

Many analysts enter the occupation with several years of work experience. Organizations that specialize in certain fields typically try to hire candidates who have experience in those areas. For example, tax preparation firms may prefer candidates who have worked as an accountant or auditor, and software companies might seek those with experience as a computer systems analyst.

**Related College Majors**

- Accounting
- Auditing
- Business Management & Administration
- Economics

## Fast Fact

When it comes to cybersecurity in the workplace, Generation Z employees don't seem to care very much, according to a survey by professional services firm EY. For instance, 58% of Gen Z-ers put off implementing mandatory IT improvements as long as possible, whereas only 15% of baby boomers reported doing the same. They were also most guilty of password reuse between private and business accounts, and of blindly accepting cookies. One reason for the trend evidenced by this report, it is suggested, is that individuals in the Gen Z bracket are simply over-familiar with technology, and having never been without it, take it for granted. Consequently, "there is an immediate need for organizations to restructure their security strategy with human behavior at the core"—a perfect opportunity for cybersecurity management consultants to come in and show them how it's done! Source: TechRadar.com

Photo via iStock.com/Alessandro Biascioli. [Used under license.]

- Entrepreneurial Studies
- Financial Management
- General Business
- Insurance
- International Business
- Marketing

### Adult Job Seekers

Adults seeking to become management analysts should come from a background or possess a strength in business, social science, math or finance. Schooling and/or certification is still potentially required, so thought should be given to an individual's long-term professional goals.

### Professional Certification and Licensure

The Institute of Management Consultants USA (IMC USA) offers the Certified Management Consultant (CMC) designation to those who meet minimum levels of education and experience and who complete other requirements. Management analysts are not required to get certification, but having the credential may give jobseekers a competitive advantage.

## EARNINGS AND ADVANCEMENT

The median annual wage for management analysts was $93,000 in 2021. The lowest 10 percent earned less than $50,190, and the highest 10 percent earned more than $163,760.

Management analysts working for consulting firms are usually paid a base salary in addition to a year-end bonus. Self-employed analysts are paid directly by their clients, typically by either the hour or the project.

Analysts often work many hours under tight deadlines. Some work more than 40 hours per week. As management analysts gain experience, they often take on more responsibility. Senior-level analysts may supervise teams working on complex projects and may become involved in seeking out new business. Those with exceptional skills may eventually become partners in their organization and focus on attracting new clients and bringing in revenue. Senior analysts may leave consulting and move to management positions at non-consulting organizations.

## EMPLOYMENT AND OUTLOOK

Management analysts held 950,600 jobs in 2021. Employment of management analysts is projected to grow 11 percent from 2021 to 2031, much faster than the average for all occupations.

About 101,900 openings for management analysts are projected each year, on average, over the decade. Many of those openings are expected to result from the need to replace workers who transfer to different occupations or exit the labor force, such as to retire.

Demand for consulting services is expected to increase as organizations seek ways to improve efficiency and control costs. As markets become more competitive, firms will need to use resources more efficiently.

Information technology (IT) consultants are expected to see high demand. Businesses will seek out consulting firms to help them attain a high level of cybersecurity and make sure their IT systems are efficient and up to date.

Growth is expected to be particularly strong in smaller consulting companies that specialize in specific industries or types of business function, such as information technology or human resources. Government agencies also are expected to seek the services of management analysts as they look for ways to reduce spending and improve efficiency.

## Related Occupations

- Accountant/Auditor
- Administrative Services/Facilities Manager
- Budget Analyst
- Cost Estimator
- Economist
- Financial Analyst
- Financial Manager
- Market Research Analyst
- Operations Research Analyst

## MORE INFORMATION

**Global Academy of Finance and Management (GAFM)**
P.O. Box 293
1670F E. Cheyenne Mountain Boulevard
Colorado Springs, CO 80906
www.gafm.com

**Institute of Management Accountants (IMA)**
10 Paragon Drive, Suite 1
Montvale, NJ 07645-1760
800.638.4427
ima@imanet.org
www.imanet.org

**Institute of Management Consultants (IMC)**
2598 E. Sunrise Boulevard, Suite 2104
Fort Lauderdale, FL 33304
800.793.4992
www.imcusa.org

*Daniel Vest*

## *Conversation With...*
## TOM KENNEDY

The Kennedy Group, Boston, Massachusetts
Management consultant, 15 years

**What was your individual career path in terms of education/training, entry-level job, or other significant opportunity?**

I founded The Kennedy Group about 15 years ago. We focus on communications skills: presentation skills, public speaking, media skills. I had spent 20 years in broadcast news, doing television news, radio talk; I was a program director. In the mid-90s, I was doing talk at WBZ radio in Boston when it went all news and I had to reinvent myself. A dear friend said, "You should do what I'm doing—coach business executives." From my years in broadcast, I had the communication skills, but not the business acumen. I spent time in the mid-1990s building a business consultancy. Coincidentally, my wife was get-ting her MBA and I, vicariously through her, studied business. I'm also involved with the Institute of Management Consultants, which is a national organization. Most of our members are independent consultants.

**What are the most important skills and/or qualities for someone in your profession?**

Interpersonal skills are the essential skill of any management consultant practice because you're selling something intangible—you're selling strategy. You have to make clients comfortable in the belief that you're acting in their best interest. Consultancy is in a very real sense a fiduciary role. Your goal is always your client's success.

The ability to help clients identify the problem is also an important skill. The problem that clients bring to a consultant is rarely the actual problem. They may say the problem is that revenues are down or they're having a problem finding the right people. But if revenues are down, it's because there's something broken. If they can't attract good people, something's wrong with the corporate culture. You need to ask the right questions and listen. Until we truly understand the underlying issues, we can't get to the crux of the problem.

We all bring different communication styles to the table. In my business, the communication business, bridging those styles is essential

**What do you wish you had known going into this profession?**

That I was becoming a sales person. You have to go out there and sell yourself, get clients and get business. A lot of people, especially the more technical people, are not comfortable with that at all.

**Are there many job opportunities in your profession? In what specific areas?**
The beauty of consulting is there are consultants in every conceivable area of business.

Independent management consultants, by virtue of the fact that they're independent, do very little hiring. The big guys, the McKinsey & Companys, the Deloittes, usually hire MBAs, which is a good opportunity for a younger person to get to know the business—if they go in with the idea that they are there to learn, not to teach. They are the fact finders, they bring the data to the "gray hairs," so to speak. Many management consultants try to be all things to all people. My advice is pick a niche and be the best there is at it and you'll succeed.

**How do you see your profession changing in the next five years? What role will technology play in those changes, and what skills will be required?**
Technology of course changes everything. There's something new every 18 months or two years. Marketing is the best example. Social media has really changed the field of marketing. Why anyone would buy a TV spot again, when most people DVR and fast forward through commercials?

You have to be video-confident—camera-ready, I call it—in business today. In all areas of management consulting, technology will drive innovation, but it will never replace the people who have been there, done that. Technology is just a tool to leverage the wisdom, the knowledge and the benefit that the consultant brings

**What do you like most about your job? What do you like least about your job?**
What I enjoy most are the brilliant people I work with on the cutting edge of business and law and technology. That's the fun part, meeting cutting-edge people doing fascinating things, changing the world, literally changing the world. I learn more from my clients than they do from me.

What I enjoy least is very minor. It's stuff like paperwork. I hate doing the invoicing, the record-keeping, the financial stuff, so I farm that out. If there's anything substantial that anyone hates about their job, they're in the wrong job.

**Can you suggest a valuable "try this" for students considering a career in your profession?**
An internship at a consulting firm is certainly worth considering. Go out to client consultancies and taken it all in. Have a plan, but be prepared to throw the plan away. Have a mentor. Have a person whose opinion you can trust.

*This interview was originally published in 2015.*

# Mathematician/ Statistician

## Snapshot

**Career Cluster(s):** Science, Technology, Engineering & Mathematics

**Interests:** Problem Solving; Analyzing Data; Communicating with Others

**Earnings (Yearly Average):** $96,280

**Employment & Outlook:** Much Faster Than Average Growth Expected

## OVERVIEW

### Sphere of Work

Mathematicians and statisticians analyze data and apply computational techniques to solve problems. They typically work in any field that benefits from data analysis, including education, government, healthcare, and research and development. In the latter field, they may work on teams attempting to advance technology such as the latest cybersecurity measures.

### Work Environment

Photo via iStock.com/metamorworks. [Used under license.]

Mathematicians and statisticians typically work in offices. A large number of these professionals work for the government and in scientific research. They also may work on teams with engineers, scientists, and other specialists related to their field, including on projects related to burgeoning technology. Mathematicians and statisticians who work in academia also spend their time outside of the classroom at research facilities.

### Profile

**Working Conditions:** Inside
**Physical Strength:** Light Work
**Education Needs:** Bachelor's Degree; Master's Degree; Doctoral Degree
**Licensure/Certification:** Not Required
**Opportunities for Experience:** Internship; Part-Time Work
**Interest Score:** ICA

## Occupation Interest

Mathematicians and statisticians are well-versed in various branches of mathematics and usually have an analytical mind that is attuned to problem solving. It takes both patience and diligence to become a mathematician or statistician. Generally, these intelligent professionals are good at working with complex data and are highly capable of communicating their findings and ideas to others.

## A Day in the Life—Duties and Responsibilities

To solve problems, mathematicians rely on statisticians to design surveys, questionnaires, experiments, and opinion polls for collecting the data they need. For most surveys and opinion polls, statisticians gather data from some people in a particular group. Statisticians determine the type and size of this sample for collecting data in the survey or poll.

Following data collection is analysis, which involves mathematicians and statisticians using specialized statistical software. In their analyses, mathematicians and statisticians identify trends and relationships within the data. They also conduct tests to determine the data's validity and to account for possible errors. Some help write software code to analyze data more accurately and efficiently.

Mathematicians and statisticians present findings from their analyses and discuss the data's limitations to ensure accurate interpretation. They may present written reports, tables, and charts to team members, clients, and other users.

Typically, mathematicians and statisticians work on teams with other specialists to solve problems.

### Duties and Responsibilities

- Deciding what data are needed to answer specific questions or problems
- Applying mathematical theories and techniques to solve practical problems in business, engineering, the sciences, and other fields
- Designing surveys, experiments, or opinion polls to collect data
- Developing mathematical or statistical models to analyze data
- Interpreting data and communicating analyses to technical and nontechnical audiences
- Using statistical software to analyze data and creating visualizations to aid decision making in business

## OCCUPATION SPECIALTIES

### College and University

Mathematicians and statisticians working in postsecondary schools may study theoretical or abstract concepts in these fields. They identify, research, and work to resolve unexplained issues in mathematics and explore mathematical or statistical theories to increase knowledge and understanding about the field.

### Government

Mathematicians and statisticians working in government develop surveys and collect and analyze data on a variety of topics, including employment, crop production, and energy use. At all levels of government, these data help to inform policy proposals and decisions that affect the public.

### Healthcare

Statisticians known as biostatisticians or biometricians work in pharmaceutical companies, public health agencies, or hospitals. They may design studies to test whether drugs successfully treat diseases or medical conditions. They may also help identify the sources of outbreaks of illnesses in humans and animals.

### Research and Development

Mathematicians and statisticians design experiments for product testing and development. For example, they may help design experiments to see how car engines perform when exposed to extreme weather or analyze consumer data for use in developing marketing strategies.

## WORK ENVIRONMENT

### Immediate Physical Environment

Work environments for mathematicians and statisticians are primarily office settings, laboratories, and computer facilities. Some may also work in academic and research environments. Depending on the specialized field they work in, the environment may vary.

### Human Environment

Often mathematicians and statisticians should expect to work in communicative environments. These professionals must perform well working with others to find solutions to problems and be able to verbally convey their findings in a way that both their peers and the average person without advanced knowledge of math or science can comprehend.

### Technological Environment

Mathematicians and statisticians extensively use computers in their work. In addition to practical office suite software, mathematicians use a variety of specialized data analysis and scientific programs such as the statistical analysis software SPSS Statistics, or the programming language and numeric computing software MATLAB. Taking courses in computer science during secondary, undergraduate and graduate studies can help prepare a mathematician or statistician to utilize these programs and more when they enter the workforce. Mathematicians seeking to specialize in a particular area, such as AI, must become familiar with the technology to which their mathematical skills will be applied.

## EDUCATION AND TRAINING

### High School/Secondary

Aspiring mathematicians and statisticians should take as many high school math courses as possible. Where available, they should study calculus and seek advanced-placement classes in science and math. Extracurricular activities involving technology or mathematics are also encouraged. Most high schools have math clubs. In the United States, there is a high school and two-year college mathematics honor society called Mu Alpha Theta that holds conventions and competitions for students. Students interested in becoming mathematicians should seek a wide base of understanding of the subject before enrolling in a college or university.

#### Transferable Skills and Abilities

**Analytical Skills**
- Using mathematical techniques and models to evaluate large amounts of data

**Communication Skills**
- Explaining technical concepts and solutions in nontechnical ways

**Logical-thinking Skills**
- Understanding and using computer programming languages
- Designing and developing models to analyze data

**Math Skills**
- Using statistics, calculus, and linear algebra to develop models and analyses

**Problem-solving Skills**
- Devising solutions to problems encountered in science, engineering, and other fields

**Suggested High School Subjects**
- Algebra
- Biology
- Calculus
- Chemistry
- Civics
- Economics
- English
- Entrepreneurship

- Geometry
- History
- Organic Chemistry
- Physics
- Statistics
- Trigonometry

**Related Career Pathways/Majors**

*Science, Technology, Engineering & Mathematics Cluster*

- Science & Mathematics Pathway

## Postsecondary

Mathematicians and statisticians typically require at least a master's degree in mathematics or statistics. For those interested in working towards senior mathematics or statisticians or other managerial roles, a master's or doctoral degree may be required for these advancement opportunities. However, some positions are available to those with a bachelor's degree.

For jobs with the federal government, candidates need at least a bachelor's degree in mathematics or significant coursework in mathematics. In private industry, mathematicians typically need either a master's or a doctoral degree.

Most colleges and universities have bachelor's degree programs in mathematics. Courses usually include calculus, differential equations, and linear and abstract algebra. Mathematics students also commonly take courses in a related field, such as computer science, physics, or statistics.

Many universities offer master's and doctoral degrees in theoretical or applied mathematics. Students who get a doctoral degree may work as professors of mathematics in a college or university.

Statisticians typically need a master's degree, but some entry-level positions may accept candidates with a bachelor's degree.

Statisticians typically have a degree in mathematics, economics, or a related field.

Students majoring in statistics also may take courses in another field, such as computer science, life sciences, or physical sciences. These courses may help prepare students to work in a variety of industries. For example, coursework in biology, chemistry, or health sciences is useful for testing pharmaceutical or agricultural products. Physics may be useful for statisticians working in manufacturing on quality improvement.

**Related College Majors**

- Applied Mathematics
- Biostatistics
- Computational Mathematics
- Economics
- Mathematics
- Quantitative Methods
- Research Methodology
- Statistics

## Adult Job Seekers

Adults interested in pursuing a career in mathematics or statistics may have a more difficult time qualifying for positions. Those who studied mathematics or statistics as an undergraduate may return to school to receive a master's degree in a relevant course of study. If an adult does not have an undergraduate degree in something related to math and statistics, they may consider taking continuing education courses to pursue a career in this field.

## Professional Certification and Licensure

There is no formal certificate or license that a mathematician or statistician receives upon graduating from a master's or doctoral program. However, some of the fields that employ mathematicians and statisticians require special certification. For example, teachers and actuaries (individuals who work with insurance policies, pensions, and annuities) require separate training and licensure.

## Additional Requirements

Mathematics and statistics require diligence and patience. Typically, mathematicians and statisticians thoroughly enjoy the process of problem solving and the slow gratification that comes with pursuing a solution. Those working in this field must have an immaculate understanding of the many different disciplines of mathematics as it is at the core of all their work. In addition, they must be excellent communicators and be comfortable explaining their work and findings to others.

## EARNINGS AND ADVANCEMENT

Earnings depend on the educational background of the mathematician or statistician and their years of experience in the field. Median annual earnings of mathematicians were $108,100 in 2021. The lowest 10 percent earned less than $61,760, and the highest 10 percent earned more than $169,500.

For statisticians, median annual earnings were $95,570 in 2021. The lowest 10 percent earned less than $49,350, and the highest 10 percent earned more than $157,300.

## EMPLOYMENT AND OUTLOOK

Mathematicians and statisticians held 36,100 jobs in 2021. Employment is expected to show a much faster than average change through the year 2031 with 31 percent job growth. Growth is expected to result from more widespread use of statistical analysis to inform business, healthcare, and policy decisions.

Many openings are expected to result from the need to replace workers who transfer to different occupations or exit the labor force, such as to retire. Applicants who have strong communication skills and the ability to interpret and present data and findings should have the best job prospects.

The amount of digitally stored data will increase over the next decade as more people and companies conduct business online and use social media, smart phones, and other mobile devices. As a result, businesses will increasingly need mathematicians and statisticians to analyze the large amount of information and data collected. Statistical analyses will help companies improve their business processes, design and develop new products, and advertise products to potential customers.

Many of the new jobs for mathematicians and statisticians are expected to be in academia, research and development, consulting, and computer systems design and related services. Many of the positions available in these industries require knowledge in related scientific branches of study in addition to mathematics and statistics. One of the most common fields available to professional mathematicians and statisticians is scientific and technical services. Another very commonly occupied field is government. In this field, statisticians are more concerned with data on population and other figures regarding the health and environment of the population rather than data on various scientific findings.

## Famous First

**British mathematician Alan Turing (1912-1954) is frequently cited as the father of modern computing science. His idea for a universal computing machine (also called a Turing machine) that would be able to decode and perform any set of instructions, provided a foundation for the architecture of modern computers.** Source: mentalfloss.com

Photo via Wikimedia Commons. [Public domain.]

### Related Occupations

- Actuary
- Computer Programmer
- Computer Systems Analyst
- Economist
- Financial Analyst
- Market Research Analyst
- Operations Research Analyst
- Physicist/Astronomer
- Postsecondary Teacher
- Survey Researcher

## MORE INFORMATION

**American Mathematical Society (AMS)**
201 Charles Street
Providence, RI 02904-2213
800.321.4267
www.ams.org

**American Statistical Association (ASA)**
732 North Washington Street
Alexandria, VA 22314-1943
888.231.3473
asainfo@amstat.org
www.amstat.org

**Center for Statistics and Applications in Forensic Evidence (CSAFE)**
195 Durham Center
613 Morrill Road
Ames, IA 50011
www.forensicstats.org

**Society for Industrial and Applied Mathematics (SIAM)**
3600 Market Street, 6th Floor
Philadelphia, PA 19104
800.447.7426
www.siam.org

**U.S. Office of Personnel Management (OPM)**
1900 E Street, NW
Washington, DC 20415-1000
202.606.1800
www.opm.gov

---

*Emily Porter and Stuart Paterson; updated by Daniel Vest*

## *Conversation With...*

# KRISTIN LAUTER

Principal Researcher & Research Manager
Microsoft Research, Cryptography Group, Redmond WA
Mathematician, 25 years

**What was your individual career path in terms of education/training, entry-level job, or other significant opportunity?**

I graduated from high school when I was 15 and started college at the University of Chicago. I did my undergraduate degree in Mathematics and continued there until I earned my PhD in Mathematics in 1996. The University of Chicago is a very theoretical place and doesn't even have an engineering school, so it wasn't obvious that I would end up in industry. My first job was as an Assistant Professor of Mathematics at the University of Michigan. I taught number theory and coding theory and was fascinated by my engineering graduate students and their experiences in industry. After three years, curiosity got the better of me and I accepted a job with Microsoft Research in cryptography and coding theory. Cryptography is the science of keeping secrets and forms the foundation for secure internet commerce. Coding theory is the theory of error correcting codes, which protects digital information to make it resistant to error. So, for instance, if a CD gets a scratch, it can still be played. I've been at Microsoft for 16 years, managing the Cryptography Research team for eight.

I currently serve as president of the Association for Women in Mathematics, an organization devoted to advancing women and girls in mathematics.

**What are the most important skills and/or qualities for someone in your profession?**

Love of mathematics is the number one thing. Problem-solving skills are important. Mathematics helps you hone your thinking skills and that goes both ways—being a good thinker helps you with mathematics. You also have to be able to communicate, verbally and in writing, the importance of mathematics and the importance of your work. An important skill for mathematicians that gets overlooked is the ability to write code.

**What do you wish you had known going into this profession?**

I had not realized how few women there would be at the top levels of the profession. As a woman, it's important to encourage other young women and girls in the profession.

**Are there many job opportunities in your profession? In what specific areas?**
The thinking skills that you get as a mathematician allow you to do many jobs. If you have a bachelor's in mathematics, you have a lot of options—being an accountant, doing actuarial science in the insurance industry, business administration or going into engineering of many different types. If you're interested in pursuing a career in research, when you get your PhD, you can certainly be a professor. Teaching mathematics at any level is so important for society. It's a really great profession. In the pharmaceutical industry and in the biomedical field in general, mathematical modeling is very important. I have colleagues and friends working with physicians on using mathematics to understand disease and treatments. There are jobs in my area, which is security and privacy, information theory and information technology; in the high-tech industry, both in communications and in research about the design of large data centers; and in energy. An emerging area is called Big Data, or data analytics, which is helping companies understand their data so they can provide useful services to their customers and determine things like how much inventory they need at a given time, how many people to employ, how much cloud storage they need. Another great career path for mathematicians is working for the National Security Agency (NSA), which is the largest employer of mathematicians in the country. It's all about national security ... protecting information and discovering information.

**How do you see your profession changing in the next five years? What role will technology play in those changes, and what skills will be required?**
I think there will be increased recognition of the importance of mathematics training in the industries I just mentioned. I see an expansion of opportunities for students with mathematics training, but that will be accompanied by the need for them to be fluent in computer languages.

**What do you enjoy most about your job? What do you enjoy least about your job?**
I enjoy talking to people about mathematics and I enjoy learning new problems that mathematics can be used to solve. For example, today I spoke with the head of the bioinformatics and genomics center at Stanford University about how they can get access to patient data to work on cures. The work I do helps protect the privacy of information through encryption while still allowing you to perform computations on that data.

I now have a lot of management and administrative responsibilities. That's not my favorite part of the job.

**Can you suggest a valuable "try this" for students considering a career in your profession?**
Try solving puzzles. If you get a puzzle book and find you enjoy it, mathematics will be a good fit. Try writing computer programs, either by taking a class or using online resources. Writing a computer program is like writing an algorithm to solve a problem, so it's a good proxy for what a mathematician does.

*This interview was originally published in 2015.*

# Medical Records Specialist

### Snapshot

**Career Cluster(s):** Business Management & Administration; Human Services

**Interests:** Accounting; Data; Health; Recordkeeping

**Earnings (Yearly Average):** $46,660

**Employment & Outlook:** As Fast As Average Growth Expected

## OVERVIEW

### Sphere of Work

Medical records specialists compile, process, and maintain patient files. They also may classify and enter patients' medical information into the healthcare industry's numerical coding system. The confidentiality and security of protected health information is of paramount importance to the job, thus one aspect of a medical records specialist's job is making sure patient records are safe from data breaches.

### Work Environment

Medical records specialists typically work at a computer.

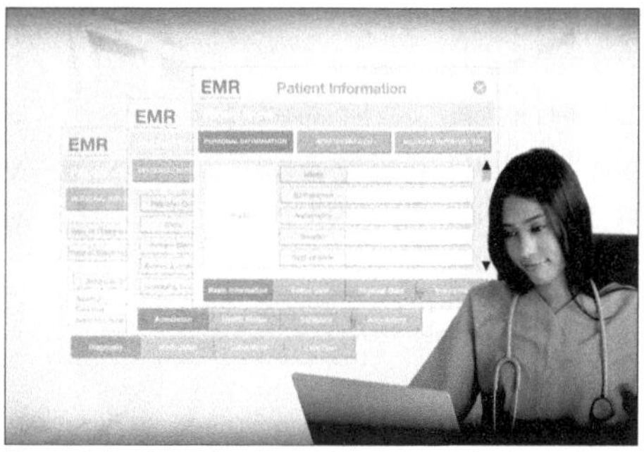

Photo via iStock.com/pandpstock001. [Used under license.]

Most medical records specialists work full time. In healthcare facilities that are always open, such as hospitals, specialists may work shifts, including nights or weekends.

## Occupation Interest

This is one of the few health-related occupations in which there is no direct hands-on patient care. However, a desire to be of service is often what drives people to enter this profession, and as their job is a crucial part of the healthcare services chain, people in this role will be conscientious, reliable and disciplined.

### Profile

**Working Conditions:** Inside
**Education Needs:** Postsecondary nondegree award
**Licensure/Certification:** Various
**Interest Score:** CES

## A Day in the Life—Duties and Responsibilities

Medical records specialists have a variety of data entry and recordkeeping tasks. They may gather patients' medical histories, symptoms, test results, treatments, and other health information and enter the details into electronic health records (EHR) systems. Some workers categorize medical information for purposes such as insurance reimbursement and providing data to clinicians.

When handling medical records, these workers follow administrative, ethical, and legal requirements for safeguarding patient privacy. Medical records specialists also may serve as gatekeepers for access to patient files. They ensure access only to authorized people and retrieve, scan, and transmit files according to established protocols.

Medical coders assign the diagnosis and procedure codes for patient care, population health statistics, and billing purposes. For example, they might review patient information for preexisting conditions, such as diabetes, to ensure proper coding of patient data. They also work as the liaison between healthcare providers and billing offices.

Although medical records specialists do not provide direct patient care, they work regularly with registered nurses and other healthcare workers. They meet with these workers to clarify diagnoses or to get additional information.

### Duties and Responsibilities

- Reviewing patients' records for timeliness, completeness, and accuracy
- Using classification systems to assign clinical codes for patients' diagnoses, procedures, medical services, and related information
- Maintaining and retrieving records for insurance reimbursement and data analysis
- Electronically recording data for collection, storage, analysis, retrieval, and reporting
- Ensuring confidentiality of patients' records

## WORK ENVIRONMENT

### Immediate Physical Environment

Most medical records and health information technicians work in a medical office environment, with much of their time spent at a desk, behind a computer screen.

### Human Environment

Medical records specialists will routinely work with other health professionals, and their communication skills must be strong as they confer with physicians and financial personnel regarding patient's sensitive medical information.

### Technological Environment

Various software program types will be used in the course of this work, including but not limited to medical, database, and categorization/classification software.

## EDUCATION AND TRAINING

### High School/Secondary

High school students may benefit from taking classes in subjects such as biology, computer science, and anatomy.

**Suggested High School Subjects**

- Anatomy
- Biology
- Computer Science
- Earth or Life or Physical Science
- English

**Related Career Pathways/Majors**

*Health Science Career Cluster*

- Health Informatics Pathway

*Human Services Career Cluster*

- Consumer Services Pathway

## Postsecondary

Medical records specialists typically need a postsecondary certificate to enter the occupation, although some qualify with a high school diploma. Others might need an associate's degree. Certification may be required or preferred.

A high school diploma or equivalent and experience in a healthcare setting are enough to qualify for some positions, but others may require a postsecondary certificate or an associate's or bachelor's degree.

Community colleges and technical schools offer certificate and associate's degree programs for medical records specialists. These programs typically include courses in medical terminology, health data requirements and standards, and classification and coding systems.

### Transferable Skills and Abilities

**Analytical Skills**
- Interpreting medical documentation to assess diagnoses, which they then code into a patient's medical record

**Detail-oriented**
- Being precise about verifying and coding patient information

**Integrity**
- Exercising discretion and acting ethically when working with patient data
- Protecting patient confidentiality, as required by law

**Interpersonal Skills**
- Discussing patient information, discrepancies, and data requirements with physicians, finance personnel, and other workers involved in patient care and recordkeeping

### Related College Majors

- Administration
- Computer Science
- Records & Information Management

## Professional Certification and Licensure

Employers may prefer to hire medical records specialists who have certification, or they may expect applicants to earn certification shortly after being hired. For example, some medical records specialists earn the Registered Health Information Tech-

### Famous First

**In the 1960s, Electronic Health Records (EHR) made their debut at the Mayo Clinic. At first considered a luxury, it was only the biggest hospitals that could afford to operate an EHR system. By 2004, President George W. Bush would create the Office of the National Coordinator for Health Information Technology, which promised access to EHRs for every American within 10 years, as EHRs had become a widespread and necessary component of medical care.**
Source: nethealth.org

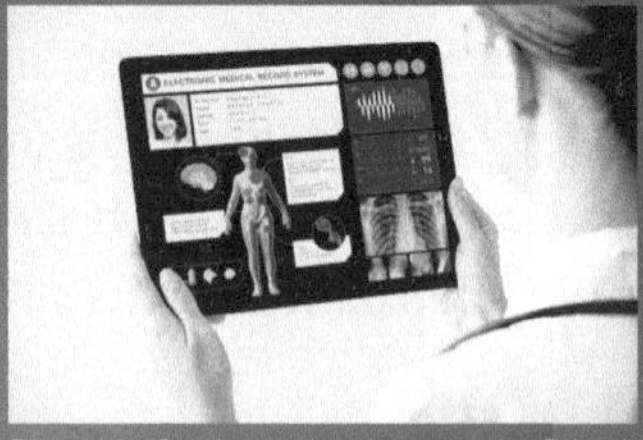
Photo via iStock.com/metamorworks. [Used under license.]

nician (RHIT) credential; certifications for medical coders include the Certified Billing & Coding Specialists (CBCS), Certified Coding Associate (CCA), Certified Coding Specialist (CCS), and Certified Professional Coder (CPC).

Certifications usually require candidates to pass an exam and might require previous experience or education. Certificate, associate's, or bachelor's degree programs may help students to meet these requirements.

## EARNINGS AND ADVANCEMENT

The median annual wage for medical records specialists was $46,660 in 2021. The lowest 10 percent earned less than $29,430, and the highest 10 percent earned more than $74,200.

Medical records specialists may advance to become health information technologists or medical registrars or medical or health services managers after completing a higher certification program or earning a degree in health information technology. Requirements vary by facility.

## EMPLOYMENT AND OUTLOOK

Medical records specialists held about 186,400 jobs in 2021. Employment is projected to grow 7 percent from 2021 to 2031, about as fast as the average for all occupations.

About 14,900 openings for medical records specialists are projected each year, on average, over the decade. Many of those openings are expected to result from the need to replace workers who transfer to different occupations or exit the labor force, such as to retire.

An increasing share of the population is entering old age, and this group typically requires more medical services. As a result, more medical records specialists will be needed to convert related health information into standardized codes to be used for insurance reimbursement and other purposes.

Additional records, coupled with widespread use of electronic health records (EHRs) by healthcare providers, will support demand for specialists to code and maintain the associated information in all areas of the healthcare industry.

### Related Occupations

- Health Information Technologist/Medical Registrar
- Information Clerk
- Medical/Health Services Manager
- Medical Assistant
- Medical Transcriptionist
- Pharmacy Technician

## MORE INFORMATION

**American Health Information Management Association (AHIMA)**
233 N. Michigan Avenue, 21st Floor
Chicago, IL 60601
800.335.5535
info@ahima.org
www.ahima.org

**American Academy of Professional Coders (AAPC)**
2233 S Presidents Drive Suite F
Salt Lake City, UT 84120
800.626.2633
www.aapc.com

**Commission on Accreditation for Health Informatics and Information Management Education (CAHIIM)**
200 East Randolph Street, Suite 5100
Chicago, IL 60601
312.235.3255
info@cahiim.org
www.cahiim.org

**National Healthcareer Association (NHA)**
11161 Overbrook Road
Leawood, KS 66211
800.499.9092
www.nhanow.com

*Daniel Vest*

# Network and Computer Systems Administrator

## Snapshot

**Career Cluster(s):** Business Management & Administration; Information Technology; Science, Technology, Engineering & Mathematics

**Interests:** Computer Science; Data; Statistics; Technology

**Earnings (Yearly Average):** $80,600

**Employment & Outlook:** Slower Than Average Growth Expected

## OVERVIEW

### Sphere of Work

Computer networks are critical parts of almost every organization. Network and computer systems administrators are responsible for the day-to-day operation of these networks. They organize, install, and support an organization's computer systems, including local area networks (LANs), wide area networks (WANs), network segments, intranets, and other data communication systems. Protecting these systems from cyberattacks is thus a top-of-mind issue for people in this profession, and they must regularly be guard for breaches of security, allowing operations to continue running smoothly.

Photo via iStock.com/shironosov. [Used under license.]

### Work Environment

Although many network and computer systems administrators are employed by firms in the computer systems design and related services industry, they work in a variety of settings. Some might administer systems and networks for financial firms, and others work in hospitals or local government offices.

### Profile

**Working Conditions:** Inside
**Education Needs:** Bachelor's Degree
**Licensure/Certification:** Vendor certifications
**Opportunities for Experience:** Postsecondary degree; Certificate
**Interest Score:** IRC

Network and computer systems administrators work with many types of workers, including other IT workers, such as computer support specialists, database administrators, computer network architects, and computer and information systems managers.

## Occupation Interest

Network and computer systems administrators are investigative types, meaning they love working with ideas, and excel in tasks that require an extensive amount of thinking. Searching for facts and figuring out problems mentally are where they shine, thus in the ever-possible case of cyberattacks, they are the first line of defense.

## A Day in the Life—Duties and Responsibilities

Administrators manage an organization's servers and desktop and mobile equipment. They ensure that email and data storage networks work properly. They also make sure that employees' workstations are working efficiently and stay connected to the central computer network. Some administrators manage telecommunication networks.

Administrators may help network architects design and analyze network models. They also participate in decisions about buying future hardware or software to upgrade their organization's network. Some administrators provide technical support to computer users, and they also may supervise computer support specialists who help solve users' problems.

### Duties and Responsibilities

- Determining an organization's system needs and installing network hardware and software
- Making needed upgrades and repairs to networks and ensuring that systems are operating correctly
- Maintaining network and computer system security
- Evaluating and optimizing network or system performance
- Adding users to a network, and assigning and updating security permissions on the network
- Training users in the proper use of hardware and software
- Interpreting and solving problems when a user or an automated monitoring system alerts them that a problem exists

## WORK ENVIRONMENT

### Immediate Physical Environment

Network and computer systems administrators work in office environments, generally, spending long hours sitting at a desk behind a computer screen.

### Human Environment

Network and computer systems administrators will do best to have strong communication skills, as they do not work in a bubble: some are responsible for the training of other administrators, as well needing to communicate to other professionals in the management chain. Network and computer administrators will sometimes need to train others in the use of various software and on how hardware works as well.

### Technological Environment

By the nature of their job, network and computer systems administrators work in a highly technological environment. They possess a range of knowledge of various hardware and software, are frequently learning new things, as well as making evaluations of and performing updates on the networks they oversee.

## EDUCATION AND TRAINING

### High School/Secondary

High school students interested in becoming network and computer systems administrators should focus on computer courses, as well as math-related courses. Becoming one sometimes requires a bachelor's degree, so students should prepare themselves for postsecondary education, including researching the schools they would like to consider attending to achieve that goal.

**Suggested High School Subjects**

- Algebra
- Applied Mathematics
- Computer Science
- Geometry
- Pre-Calculus
- Statistics

**Related Career Pathways/Majors**

*Business Management & Administration Career Cluster*

- General Management Pathway
- Operations Management Pathway

*Information Technology Career Cluster*

- Network Systems Pathway

## Transferable Skills and Abilities

**Analytical Skills**

- Evaluating networks and systems to make sure that they perform reliably
- Anticipating new requirements as customers' needs change

**Communication Skills**

- Describing problems and their solutions to non-IT workers

**Multitasking Skills**

- Working on many problems and tasks at the same time

**Problem-Solving Skills**

- Quickly resolving problems that arise with computer networks

## Postsecondary

Most employers require network and computer systems administrators to have a bachelor's degree in a field related to computer or information science. Others may require only a postsecondary certificate or an associate's degree.

Although some employers require a postsecondary certificate or an associate's degree, most require network and computer systems administrators to have a bachelor's degree in a related field, such as computer and information technology. There are degree programs that focus on computer network and system administration. However, because administrators work with computer hardware and equipment, a degree in computer engineering or electrical engineering usually is acceptable as well. Programs in these fields frequently include classes in computer programming, networking, or systems design.

Because network technology is constantly changing, administrators need to keep up with the latest developments. Many continue to take courses throughout their careers and attend information technology (IT) conferences to keep up with the latest technology. Some businesses require that administrators have a master's degree.

**Related College Majors**

- Computer Engineering
- Computer Science
- General Business
- Information Systems Management
- Mathematics

### Adult Job Seekers

Adults seeking to become network and computer systems administrators should have a background or strength in computers, math or science. Some schooling may still be required, such as certificate program or an associate's degree.

### Professional Certification and Licensure

Companies generally require their network and computer systems administrators to be certified in the products they use. Certification programs usually are offered directly from vendors or from vendor-neutral certification providers. Certification validates the knowledge and the use of best practices that are required of network and computer systems administrators. Microsoft and Cisco offer some of the most common certifications.

## EARNINGS AND ADVANCEMENT

The median annual wage for network and computer systems administrators was $80,600 in 2021. The lowest 10 percent earned less than $49,560, and the highest 10 percent earned more than $130,830.

Most network and computer systems administrators work full time. Organizations depend on their computer networks, so administrators may need to work overtime to ensure that the networks are operating properly around the clock. Network administrators can advance to become computer network architects. They can also advance to managerial jobs in information technology (IT) departments, such as computer and information systems managers.

## EMPLOYMENT AND OUTLOOK

Network and computer systems administrators held 333,200 jobs in 2021. Employment is projected to grow 3 percent from 2021 to 2031, slower than the average for all occupations.

Despite limited employment growth, about 23,900 openings for network and computer systems administrators are projected each year, on average, over the decade. Most of those openings are expected to result from the need to replace workers who transfer to different occupations or exit the labor force, such as to retire.

## Fast Fact

**High on the list of tools every admin should know and use are "pentesting" tools (a.k.a. penetration testing tools), which mimic various types of attacks on your devices to see if they can break through your network defenses. Many companies employ pentesters, or so-called ethical hackers, to run these types of tasks regularly.** Source: techrepublic.com

Photo via iStock.com/GCShutter. [Used under license.]

Demand for network and computer systems administrators should continue, along with firms' investment in newer, faster technology and mobile networks. The continued expansion of cloud computing and the demand for upgraded computer equipment and software will support a need for network and computer systems administrators to maintain these systems.

### Related Occupations

- Computer/Information Systems Manager
- Computer Hardware Engineer
- Computer Network Architect
- Computer Programmer
- Computer Support Specialist
- Computer Systems Analyst
- Database Administrator/Architect
- Electrical/Electronics Engineer
- Information Security Analyst
- Software Developer/Quality Assurance Analyst/Tester
- Web Developer/Digital Designer

## MORE INFORMATION

**Association for Computing Machinery (ACM)**
1601 Broadway, 10th Floor
New York, NY 10019
212.869.7440
acmhelp@acm.org
www.acm.org

**Computing Research Association (CRA)**
1828 L Street, NW, Suite 800
Washington, DC 20036
202.234.2111
info@cra.org
www.cra.org

**IEEE Computer Society**
2001 L Street NW, Suite 700
Washington, DC 20036
202.371.0101
help@computer.org
www.computer.org

**Institute for Operations Research and the Management Sciences (INFORMS)**
5521 Research Park Drive, Suite 200
Catonsville, MD 21228
800.446.3676
informs@informs.org
www.informs.org

**National Center for Women & Information Technology (NCWIT)**
1909 26th Street, 2nd Floor
Boulder, CO 80302
303.735.6671
info@ncwit.org
www.ncwit.org

---

*Daniel Vest*

# *Conversation With...*
# DAMIAN DAVIS

Network Operations Engineer, G3 Technologies, Inc.
Chantilly, Virginia
Network administrator, 3 years

**What was your individual career path in terms of education/training, entry-level job, or other significant opportunity?**

I always wanted to do something sports-related, so my degree is in sports management with a minor in business administration. As an intern for the Washington Wizards, I got to be part of the process of evaluating a free agent. That was really cool. But those moments were few and far between, and most of the work seemed repetitive. The demand for jobs in that field was very high and, apart from trainers and deal-makers, the salaries were maybe $30,000 a year to start. Meantime, I had friends who were computer science majors asking for my help on projects.

I worked for Hertz Rent a Car for five years after graduating, but it's a high turnover industry and I realized I wasn't making enough money and wasn't seeing my family. I started looking at what fields would be a better fit. A customer worked in IT, and I already knew how to do some of the things he talked about needing to learn.

I took a full year out of work to study how to become a network engineer. I started by taking a course at a small IT school for the Cisco Certified Network Associate (CCNA) certification; it was too much to learn in a short period of time. After that class ended, I offered to volunteer at local IT firms and studied on my own. At one company in particular, the CEO and one of his directors spent an hour speaking with me. They couldn't let me volunteer due to contract specifications with clients, but the CEO saw a lot of himself in me. He picked up a huge stack of routers and switches and said, "We're about to throw these out. Why don't you take them? You can practice." That really helped me out. It pushed my development and learning, and I realized the limitations of software.

A year after making my career change, I was hired as a Junior IT Network Administrator for G3. Besides my CCNA certification, I am Network+ certified and am studying for the Cisco Certified Network Professional (CCNP) and Certified Ethical Hacker (CEH) certifications.

I love what I do.

**What are the most important skills and/or qualities for someone in your profession?**
Be willing to learn. Every day. What you learned last year is already obsolete. Also, I recommend that anyone taking an entry-level job work for a small- to mid-size company. The bigger companies will hire you for a specific duty and you'll learn how to do that very well, but not much else. Smaller companies need employees to fill a lot of shoes, so you'll have the opportunity to learn a lot very quickly. In just three years of working, I've learned SharePoint Administration, Active Directory, CyberSecurity, and Linux Ubuntu, in addition to my Cisco training and studies.

**What do you wish you had known going into this profession?**
I wish I had started with an easier certification because Cisco equipment is very difficult to learn. The A+ and Network+ certifications are foundational training—but easier—that you can always fall back on. By not starting with those certifications, I encountered situations in my first IT job that should've been easy to resolve but weren't, because I did not have the basics.

**Are there many job opportunities in your profession? In what specific areas?**
Many. The best areas for network engineers are in California, New York City, and Washington D.C.

**How do you see your profession changing in the next five years, what role will technology play in those changes, and what skills will be required?**
The way technology keeps advancing, network administrators will always be essential. More and more companies will seek out people with my skills and knowledge. A strong understanding of Windows and Linux OS, Microsoft Office, and networking principles will be essential.

**What do you enjoy most about your job? What do you enjoy least about your job?**
I most enjoy interacting with my co-workers. When I've resolved all outstanding trouble tickets, I go around the office and ask my colleagues how their day has been. This usually creates better work relationships and also can remind them of IT issues or concerns they've been having but have forgotten to tell me about. What I enjoy least is occasionally having to speak to a co-worker about inappropriate activities on their computer, like porn. It's never an easy conversation, and always awkward for all parties involved.

**Can you suggest a valuable "try this" for students considering a career in your profession?**
Apply for an internship before you graduate. You'll learn a lot of the foundational knowledge you'll need to be successful and, more importantly, you'll learn whether the career is really for you. If you're a hard worker and you like the work, chances are that company you're interning with will offer you a job.

*This interview was originally published in 2016.*

## *Conversation With...*

# PATRICIA A. VEON

Second-tier Technical Service Support
Frederick MD Country Schools, 9 years
Tech Services field, 25 years

**What was your individual career path in terms of education/training, entry-level job, or other significant opportunity?**

I came from a large family with limited resources, so higher education was not something I ever envisioned that I would be able to achieve. In high school, I took classes that would lead to basic employment, such as typing, shorthand and bookkeeping.

My first job in high school was working as a cashier. This led to a bookkeeping position.

When I lost the bookkeeping job as a young, newly-separated mother, due to sexual harassment, I realized, as I looked for work, that I could not identify my profession. Time was moving ahead, but,with only a high school education, I was falling behind.

I applied to numerous places for work, without success. Finally, at a job interview with a friend of my mother, she told me to get some training and come back. So I did.

I was very good at math, and I knew I had a talent for fixing things and solving problems. I needed to find how to make a living doing this. At the time, the government had a program to train people while they were on unemployment. I went to the job training agency in my hometown, and they gave me a test to determine my strengths and weaknesses. Sure enough, I was good at working with people and solving problems.

They helped me find a school that offered a one-year, full time, intense electronic education. By the end of the school year, I had seven job offers.

After I went to work as an electronics tech—and then as a technical writer—I started taking classes to fill in my degree. Working full-time and raising my three children, it took more than five years to complete an AA degree in Electronics from Frederick Community College in Maryland.

I learned most of my computer and network skills on the job or by studying and taking classes. I worked for Packard BioScience Company, who sent me to many classes to learn how to be a trainer. I did technical training for eight years for all of their service engineers and customers.

I moved to doing tech support in a school—even though it paid about a third of what I was making—because I was traveling all over the country and the world doing training and I wanted to get off the road. Now, I have a territory of nine schools. When the school-based tech runs into a problem they don't know how to fix, they call me. I do some network work, a lot of software work and training

**What are the most important skills and/or qualities for someone in your profession?**
You need an inherent ability to figure things out. If you're not curious about how things work, I don't think you'll make it.

Communication is vital—the most important part of my job—whether I am talking to a user to explain something on their level, or documenting a process.

Also, the tech field is always changing, making it challenging to stay up with everything. But it is never boring. If you're in the tech field, you can't stop learning. It's a job with homework.

**What do you wish you had known going into this profession?**
I wish I had known that having a higher education degree would have made me much more marketable. I also wish I had realized that I could make a living at something that I enjoyed at an earlier age. I wasted many years just making a living.

**Are there many job opportunities in your profession? In what specific areas?**
Job opportunities abound in the tech field. There is always a need for people who can set up and run networks for companies and government agencies.

**How do you see your profession changing in the next five years? What role will technology play in those changes, and what skills will be required?**
Network security is the biggest area of growth right now. If you can work in network security, you can name your price.

**What do you like most about your job? What do you like least about your job?**
I enjoy the people. I don't enjoy the office politics.

**Can you suggest a valuable "try this" for students considering a career in your profession?**
You need to find a problem and try to figure it out. I can fix anything: give me a toaster and I can either fix it, or find out how.

*This interview was originally published in 2015.*

# Operations Research Analyst

## Snapshot

**Career Cluster(s):** Business Management & Administration; Finance; Information Technology; Law, Public Safety, Corrections & Security; Science, Technology, Engineering & Mathematics

**Interests:** Accounting; Finance; Data; Mathematics; Statistics; Recordkeeping; Investigating

**Earnings (Yearly Average):** $82,360

**Employment & Outlook:** Much Faster Than Average Growth Expected

## OVERVIEW

### Sphere of Work

Operations research analysts use mathematics and logic to help organizations make informed decisions and solve problems. In terms of cybersecurity, these professionals will help organizations evaluate emerging cyber tactics, techniques, and procedures to better secure information and prevent attacks.

Photo via iStock.com/YakobchukOlena. [Used under license.]

### Work Environment

Operations research analysts spend much of their time in office settings. They may travel to gather information, observe business processes, work with clients, or attend conferences.

Some operations research analysts in the federal government work for the Department of Defense, which also employs analysts through private consulting firms.

## Profile

**Working Conditions:** Inside
**Education Needs:** Bachelor's Degree; Master's Degree
**Licensure/Certification:** Not Required
**Opportunities for Experience:** Related experience
**Interest Score:** ICE

### A Day in the Life—Duties and Responsibilities

Operations research analysts may be involved in many aspects of an organization. For example, they may help managers decide how to allocate resources, develop production schedules, oversee the supply chain, and set prices.

To begin a project, analysts first identify the problem to be solved or the processes to be improved. They typically collect data and interview clients, workers, or others involved in the business processes being examined.

Analysts then break down the problem into its various parts using statistical and database software and analytical techniques, such as forecasting and data mining. They also study the effect that different changes and circumstances would have on each of these parts. For example, to help an airline schedule flights and set ticket prices, analysts may take into account the cities involved, the amount and cost of fuel required, the expected number of passengers, the pilots' schedules, and the maintenance costs.

## Duties and Responsibilities

- Identifying problems in areas such as business, logistics, healthcare, or other fields
- Collecting and organizing information from a variety of sources, such as databases, sales histories, and customer feedback
- Gathering input from workers or subject-matter experts
- Analyzing collected data and extracting information relevant to the problem being addressed
- Developing and testing quantitative models, support software, and analytical tools
- Writing memos, reports, and other documents explaining their findings and recommendations for managers, executives, and other officials

Operations research analysts provide alternatives to pursuing different actions and may assist in achieving a consensus on how to proceed. They weigh the costs and benefits of alternative solutions or approaches in their recommendations to managers.

In terms of cybersecurity and infrastructure protection, operations research analysts are able to step in when strategic defensive decisions must be made. Some of these decisions can include physical measures such as replacing vulnerable hardware or the implementation of tamper-proof packaging. Other broader measures may include acts such as the training of employees on cybersecurity protocol.

## WORK ENVIRONMENT

### Immediate Physical Environment

Most operations research analysts work in office environments, though some travel can be involved, such as taking meeting with clients on the client's home turf, or attending conferences.

### Human Environment

Operations research analysts work alongside others in the same field to better solve problems that need addressing, so a degree of social intelligence is a necessity. These workers must also possess strong communication skills to be able to relay sometimes highly technical information in clear terms to laypersons.

### Technological Environment

Because of the nature of the job, operations research analysts do their work in a very technology-heavy environment. They use statistical and database software and analytical technique to make their assessments and implement strategies based off this research.

## EDUCATION AND TRAINING

### High School/Secondary

High school students interested in becoming operations research analysts should focus on math, statistics, and business-related courses. Entering this profession will usually require at least a bachelor's degree, if not a master's, so students should prepare themselves for postsecondary education, including researching the schools they would like to consider attending to achieve that goal.

**Suggested High School Subjects**

- Algebra
- Civics
- Computer Science
- Economics
- English

- Geometry
- History
- Physics
- Pre-Calculus
- Psychology
- Statistics
- Trigonometry

**Related Career Pathways/Majors**

*Business Management & Administration Career Cluster*

- General Management Pathway
- Operations Management Pathway

*Information Technology Career Cluster*

- Network Systems Pathway
- Information Support & Services Pathway

*Science, Technology, Engineering & Mathematics Career Cluster*

- Science & Mathematics Pathway

### Transferable Skills and Abilities

**Analytical Skills**

- Using a range of methods, including forecasting and data mining, to examine and interpret data

**Communication Skills**

- Writing memos, reports, and other documents
- Presenting their data and conclusions to managers and other
- Conveying technical information in a way that is understandable to nontechnical audiences

**Critical-thinking Skills**

- Organizing information
- Making connections between ideas and facts

**Interpersonal Skills**

- Working on teams
- Persuading managers and executives to accept their recommendations

**Math Skills**

- Using models and methods rooted in statistics, calculus, linear algebra, and other mathematics disciplines

**Problem-solving Skills**

- Diagnosing problems and studying relevant information to solve them

## Postsecondary

Operations research analysts typically need at least a bachelor's degree, but some jobs require a master's degree. Fields of degree may include operations research or a related field, such as business, mathematics, engineering, or computer science.

Because operations research is based on quantitative analysis, students need extensive coursework in mathematics. Coursework in computer science is important because analysts rely on statistical and database software to assess and model data.

Some operations research analysts are veterans of the U.S. Armed Forces. Certain positions may require applicants to undergo a background check in order to obtain a security clearance.

**Related College Majors**

- Analytics
- Computer Science

- Engineering
- Mathematics
- Physics
- Operations Research

### Adult Job Seekers

Adults seeking to become professional operations research analysts should have a background or strength in computer science or mathematics. Schooling may still be required, so thought should be given to an individual's long-term professional goals.

## EARNINGS AND ADVANCEMENT

The median annual wage for operations research analysts was $82,360 in 2021. The lowest 10 percent earned less than $48,690, and the highest 10 percent earned more than $160,850.

## EMPLOYMENT AND OUTLOOK

Operations research analysts held 104,200 jobs in 2021. Employment is projected to grow 23 percent from 2021 to 2031, much faster than the average for all occupations.

About 10,300 openings for operations research analysts are projected each year, on average, over the decade. Many of those openings are expected to result from the need to replace workers who transfer to different occupations or exit the labor force, such as to retire.

As technology advances and companies and government agencies seek efficiency and cost savings, demand for operations research analysis should continue to grow. In addition, increasing demand should occur for these workers in the field of analytics to improve business planning and decision making.

Technological advances have made it faster and easier for organizations to get data. Operations research analysts manage and evaluate data to improve business operations, supply chains, pricing models, and marketing. In addition, improvements in analytical software have made operations research more affordable and applicable to a wider range of areas. More companies are expected to employ op-

## Fast Fact

**According to Cyber magazine, Washington, D.C. is one of the best locations for work in cybersecurity because of the number of prestigious private companies with headquarters there. In 2021, there were more than 72,000 open positions in the field for applicants to try.** Source: Cybermagazine.com

Photo via iStock.com/cineman69. [Used under license.]

erations research analysts to help them turn data into information that managers use to make decisions about all aspects of their business.

### Related Occupations

- Economist
- Industrial Engineer
- Logistician
- Management Analyst
- Market Research Analyst
- Mathematician/Statistician
- Software Developer/Quality Assurance Analyst/Tester

## MORE INFORMATION

**Institute for Operations Research and the Management Sciences (INFORMS)**
5521 Research Park Drive, Suite 200
Catonsville, MD 21228
800.446.3676
informs@informs.org
www.informs.org

**Military Operations Research Society (MORS)**
2111 Wilson Boulevard, Suite 700
Arlington, VA 22201
703.933.9070
www.mors.org

*Daniel Vest*

# Personal Financial Advisor

**Snapshot**

**Career Cluster(s):** Business Management & Administration; Finance; Human Services

**Interests:** Accounting; Customer Service; Finance; Data; Statistics

**Earnings (Yearly Average):** $94,170

**Employment & Outlook:** Much Faster Than Average Growth Expected

## OVERVIEW

### Sphere of Work

Personal financial advisors provide advice on investments, insurance, mortgages, estate planning, taxes, and retirement to help individuals manage their finances. Because of the sensitive financial information they deal in, a strong awareness of cybersecurity protocols is an essential.

Photo via iStock.com/LDProd. [Used under license.]

### Work Environment

Personal financial advisors typically work in offices. Some also travel to attend conferences, teach finance seminars in the evening, and attend networking events to bring in more clients.

## Occupation Interest

One hallmark of a good personal financial advisor is that they are enterprising individuals. Passion for wealth management, as well as for customer service are key. A professional, clients-come-first attitude combined with investing acumen and market timing will propel individuals in this career forward.

### Profile

**Working Conditions:** Inside
**Education Needs:** Bachelor's Degree; Master's Degree
**Licensure/Certification:** SEC; NASAA; CFP
**Opportunities for Experience:** On-the-Job Training
**Interest Score:** ECS

## A Day in the Life—Duties and Responsibilities

Personal financial advisors assess the financial needs of individuals and help them with decisions on investments (such as stocks and bonds), tax laws, and insurance. Advisors help clients plan for short- and long-term goals, such as budgeting for education expenses and saving for retirement through investments. They invest clients' money based on the clients' decisions. Many advisors also provide tax advice or sell insurance.

Although most planners offer advice on a wide range of topics, some specialize in areas such as retirement or risk management (evaluating the investor's willingness to take chances and adjusting investments accordingly).

Many personal financial advisors spend a lot of time marketing their services, and they meet potential clients by giving seminars or participating in business and social networking.

### Duties and Responsibilities

- Meeting with clients to discuss their financial goals
- Explaining to potential clients the types of financial services they provide
- Educating clients and answering questions about investment options and potential risks
- Recommending investments to clients or select investments on their behalf
- Helping clients plan for specific circumstances, such as education or retirement
- Monitoring clients' accounts and determining if changes are needed to improve financial performance or to accommodate life changes, such as getting married or having children
- Researching investment opportunities

After financial advisors have invested funds for a client, they and the client receive regular investment reports. Advisors monitor the client's investments and usually meet with each client at least once a year to update the client on potential investments and to adjust the financial plan based on the client's circumstances or because investment options may have changed.

Many personal financial advisors are licensed to directly buy and sell financial products, such as stocks, bonds, annuities, and insurance. Depending on the agreement they have with their clients, personal financial advisors may have the client's permission to make decisions about buying and selling stocks and bonds.

## WORK ENVIRONMENT

### Immediate Physical Environment

Many personal financial advisors work in an office environment, but traveling to different locations for meetings with clients is always possible.

### Human Environment

Sociability is a key factor in the success of a personal finance advisor. This job is all about understanding the motivations and gaining the trust of the client: asking questions, getting to know what drives the customer and determining the right financial moves for them. Listening skills, the ability to process/interpret information, and relay complex idea in lay terms are important talents to hone.

### Technological Environment

Technological know-how of a personal financial advisor includes awareness of many software types, and the capability to directly buy and sell financial products, such as stocks, bonds, annuities, and insurance.

## EDUCATION AND TRAINING

### High School/Secondary

High school students interested in becoming personal financial advisors should focus on business courses as well as math and communications-related courses. Becoming a personal financial advisor typically requires a bachelor's degree, and in order to advance further in their career, a master's may be preferred, so students should prepare themselves for postsecondary education, including researching the schools they would like to consider attending to achieve those goals.

**Suggested High School Subjects**

- Civics
- Economics
- English
- History

- Pre-Calculus
- Psychology
- Statistics

**Related Career Pathways/Majors**

*Finance Career Cluster*

- Banking Services Pathway
- Securities & Investments Pathway

*Human Services Career Cluster*

- Consumer Services Pathway

## Postsecondary

Personal financial advisors typically need a bachelor's degree to enter the occupation. A master's degree and certification may improve chances for advancement.

Personal financial advisors typically need a bachelor's degree, although employers usually do not require a specific course of study. However, common fields of degree include business, social science, or mathematics. Courses in investments, taxes, estate planning, and risk management may be helpful.

After they are hired, personal financial advisors typically need on-the-job training to attain competency. During this time, new advisors work under the supervision of senior advisors and learn how to build a client network, develop investment portfolios, and perform other duties. This training usually lasts for more than a year.

**Related College Majors**

- Accounting
- Auditing
- Business Management & Administration
- Economics
- Entrepreneurial Studies
- Financial Management

### *Transferable Skills and Abilities*

**Analytical Skills**

- Assessing a range of information, including economic trends, regulatory changes, and the client's comfort with risky decisions

**Interpersonal Skills**

- Making clients feel comfortable
- Establishing trust with clients and responding well to their questions and concerns

**Math Skills**

- Working adeptly with numbers
- Determining the amount invested, how that amount has grown or decreased over time, and how a portfolio is distributed among different investments

**Sales Skills**

- Expanding their base of clients regularly
- Being convincing and persistent in selling their services

**Speaking Skills**

- Interacting with clients every day
- Explaining complex financial concepts in a way that clients understand

- General Business
- Insurance
- International Business
- Marketing

### Adult Job Seekers

Adults seeking to become personal financial advisors should have a background or strength in business, social science, math or finance. A particular educational path is not a must, but course work in investments, taxes, estate planning, and risk management are all smart choices for those interested in this profession.

### Professional Certification and Licensure

Personal financial advisors who directly buy or sell stocks, bonds, or insurance policies, or who provide specific investment advice, may need a combination of licenses that varies with the products they sell. In addition to being required to have those licenses, advisors in small firms that manage clients' investments must be registered with state regulators, and those in large firms must be registered with the U.S. Securities and Exchange Commission (SEC). Personal financial advisors who choose to sell insurance need licenses issued by state boards. Information on state licensing board requirements for registered investment advisors is available from the North American Securities Administrators Association (NASAA).

Certifications may enhance a personal financial advisor's reputation and help bring in new clients. The Certified Financial Planner Board of Standards offers the Certified Financial Planner (CFP) designation. For this certification, advisors must have a bachelor's degree, complete coursework on financial planning through a CFP Board Registered Program, have relevant work experience, pass an exam, and agree to adhere to a code of ethics.

## EARNINGS AND ADVANCEMENT

The median annual wage for personal financial advisors was $94,170 in 2021. The lowest 10 percent earned less than $47,570, and the highest 10 percent earned more than $208,000.

Personal financial advisors who work for financial services firms are often paid a salary plus bonuses. Commissions, incentive pay, and production bonuses are included in the wage data here; nonproduction bonuses are not included.

## Fast Fact

**According to the consulting firm PwC, "There are two kinds of financial services firms: those that have faced a cyberattack and those that will." Thus, investors should take care that their advisors are following cyber requirements or recommendations from financial regulators such as the SEC (i.e. encrypted e-mail messages and multi-factor authentication to access the client portal). The Financial Planning Association, a membership group of almost 21,000 advisors, has launched its own certificate in cybersecurity.** Source: CNBC

Photo via iStock.com/JuSun. [Used under license.]

Advisors who work for financial investment firms or financial planning firms or who are self-employed earn money for their services in one of two ways. They either charge a flat fee or earn commissions for the financial products that they sell.

Most personal financial advisors work full time, and some work more than 40 hours per week. They also may go to meetings on evenings and weekends to meet with existing clients or to try to bring in new ones.

A master's degree in a field such as finance or business administration may improve a personal financial advisor's chances of becoming a financial manager and of attracting new clients.

## EMPLOYMENT AND OUTLOOK

Personal financial advisors held about 330,300 jobs in 2021. Employment is projected to grow 15 percent from 2021 to 2031, much faster than the average for all occupations.

About 30,500 openings for personal financial advisors are projected each year, on average, over the decade. Many of those openings are expected to result from the need to replace workers who transfer to different occupations or exit the labor force, such as to retire.

The primary driver of employment growth will be the aging population. As large numbers of baby boomers continue to retire, they are likely to seek planning advice from personal financial advisors. Also, longer lifespans will lead to longer retirement periods, further increasing demand for financial planning services.

In addition, the replacement of traditional pension plans with individual retirement accounts is expected to continue. Many people used to receive defined pension payments in retirement, but most companies no longer offer these plans. Therefore, individuals must save and invest for their own retirement, increasing the demand for personal financial advisors.

The availability of "robo-advisors," computer programs that provide automated investment advice based on user inputs, may partially temper demand for personal financial advisors. However, the impact of this technology should be limited as consumers continue turning to human advisors for more complex and specialized investment advice over the projections decade.

### Related Occupations

- Budget Analyst
- Financial Analyst
- Financial Manager
- Insurance Sales Agent
- Insurance Underwriter
- Real Estate Broker/Sales Agent
- Securities/Commodities/Financial Services Sales Agent

## MORE INFORMATION

**Certified Financial Planner Board of Standards (CFP)**
1425 K Street NW, Unit 800
Washington, DC 20005
800.487.1497
mail@cfpboard.org
www.cfp.net

**Financial Industry Regulatory Authority (FINRA)**
1735 K Street, NW
Washington, DC 20006
301.590.6500
www.finra.org

**Global Academy of Finance and Management (GAFM)**
P.O. Box 293
1670F E. Cheyenne Mountain Boulevard
Colorado Springs, CO 80906
www.gafm.com

**North American Securities Administrators Association (NASAA)**
750 First Street NE, Suite 990
Washington, DC 20002
202.737.0999
www.nasaa.org

**U.S. Securities and Exchange Commission (SEC)**
100 F Street NE
Washington, DC 20549
800.732.0330
www.sec.gov

---

*Daniel Vest*

## *Conversation With...*

## PAUL S. VIREN

Financial Planner, Viren & Associates
Spokane, Washington
Financial planner, 24 years

**What was your individual career path in terms of education/training, entry-level job, or other significant opportunity?**

I'm a product of a liberal arts education. I graduated with degrees in psychology and religion, and I use those two skills in my practice more than I use the financial part. I did ministry work for a couple of years, but decided it wasn't for me. I got into sales and worked in a variety of positions and ended up working at a university here in Spokane in the development office. That's how I first discovered estate planning, trying to helping alumni make a bequest or a gift to the university in a legacy way. I fell in love with that whole concept. That goes back about 30 years.

I got into this profession through an insurance company, the Principal Financial Group, which had a training program that helped you get licensed and offered a lot of classes. Financial planning is a very, very difficult professional to break into because there's lots of licensing, lots of failure, lots of commission types of arrangements. After about six years, I matured out of The Principal and became an independent financial planner, doing fee-based work. Throughout that whole time, estate planning has been a cornerstone of my success and my passion for helping people in preparing for their future. I find it's a helpful starting point to motivate clients. At our first meeting, I tell them there are five documents that are universally needed: a will, power of attorney for health care, power of attorney for finances, and a living will. The fifth one is probably the most important one: a handwritten love letter to your husband or your child and other family members or friends. Usually that last item is what gets people to realize they better take care of these things.

The journey of finding a career is not always a straight line. It can take several different kinds of jobs. That was my experience until I finally found the job I love, the one I'm doing now. Along the way, I learned different skill sets that I incorporate into my work.

**What are the most important skills and/or qualities for someone in your profession?**

Being empathetic and listening are really important skills to build a relationship with a client. The profession requires a lot of technical expertise, but many people who are good technical people are horrible relationship-building people. Many times, you're dealing with people who have just gotten married or just lost a loved one or just had child or lost their child or started a business or sold their business. As people go through those chapters, that's where we enter in.

I really gravitate toward educating my clients on the world of finances. A lot of people don't know what they don't know when they go to their financial investment person. To have at least a basic understanding is important. Especially when there's a stressful point in a market cycle and there's the usual moment of panic, talking people off the edge of the roof is important, but you have to have that trust. You don't build trust by being a good technician; you build trust by being empathetic and listening.

**What do you wish you had known going into this profession?**
I wish I had known how rewarding it would be because I would have gotten into it sooner. It's so challenging to get through the first few years of this profession.

**Are there many job opportunities in your profession? In what specific areas?**
Every time I open up a professional journal, the headlines are screaming that we're all getting older and there's nobody behind us. The average age is 58 or older—it's crazy. So I think there's huge opportunity over the next 5, 10 years.

**How do you see your profession changing in the next five years? What role will technology play in those changes, and what skills will be required?**
The robo-advice that's out there is going to help some people, because we want an app on our phone. The problem with a financial planner app is it's cookie cutter. It doesn't do the empathetic listening or ask good questions. And people like experiences ... we like the Starbucks experience versus coffee out of a machine.

**What do you enjoy most about your job? What do you enjoy least about your job?**
I'm passionate about helping clients and being part of their lives and helping them achieve their goals.

We are a profession that is very, very heavily regulated. The regulation has gotten so onerous.

**Can you suggest a valuable "try this" for students considering a career in your profession?**
Get an internship with a brokerage firm or an insurance company. If nothing else, most of those places, if they're good, will help you with communication skills, which is code for selling skills. Try to find a mentor, that person who's willing to take a little bit of time to help you.

*This interview was originally published in 2017.*

# *Conversation With...*

## KYLE P. GERMAN, CRPC

Partner, Harbor Financial Group
Annapolis, Maryland
Financial planner, 18 years

**What was your individual career path in terms of education/training, entry-level job, or other significant opportunity?**

I'm a practical person. I've always been interested in finance and the movement of money. This career path has provided me with an opportunity to make a nice living doing something I enjoy. However, I knew I could take an understanding of finance with me into many fields.

So, I took business and finance courses in college, and earned a bachelor's degree from West Virginia University. These classes in no way prepared me for a career as a personal financial advisor. We spent our time studying the theory behind the numbers as opposed to the practical application.

**What are the most important skills and/or qualities for someone in your profession?**

You need to be a people person because you will spend your days meeting with individuals and families, learning about their goals, their concerns, what drives them.

You will not be micromanaged, so you need self-motivation. You need to be creative when prospecting for new clients to set yourself apart from the thousands of other smart young people trying to build a financial planning practice. So, for instance, bring coffee to a business (even if it's a construction company that starts at 5:30) so you can meet people.

Lastly, you must have a thick skin! You will be told "no" thousands of times.

**What do you wish you had known going into this profession?**

I wish had asked for help. My best advice is to offer to help an experienced advisor. This will provide you with an opportunity to learn from someone who's done it and potentially provide you with a future business partner.

**Are there many job opportunities in your profession? In what specific areas?**

Financial advisors are in demand, for two reasons:

First off, baby boomers are retiring and shifting their retirement nest eggs from company 401K savings plans to individual IRAs. This represents the largest transfer of wealth the country has ever seen. And when people retire, they need help managing their money.

Second, this profession has high turnover because it's tough to get started. At my first job, thirty other people started at the same time. Our sole responsibility was to study for the Series 7 Exam, which is basically the stockbroker's exam. We had one month to pass the test. Then we had two weeks to study and complete the Series 66—the state securities law exam—and then we moved on to the Life and Health Exam, which is for insurance. If you failed, you were out. I quickly learned the exams were the easy part. Only two of us were left by the end of the second year. Not all firms have the same exam requirements, but all will expect you to bring in new clients right away.

If you stick with it and get through the years of hard work, it's amazing how lucrative this field can be. Trust me when I say you will work harder than most for less money for ten to fifteen years (depending how successful you are) and then, once you've built your practice, you'll work less than most for more money.

**How do you see your profession changing in the next five years, what role will technology play in those changes, and what skills will be required?**
Clients, especially younger clients, want immediate online access to their statements on their smart phones. They would rather text or email than come in for a meeting. This is all based on efficiency. If you don't adapt to work with them, they'll find someone who will.

That said, in-person meetings build trust. Older clients may be dealing with life's complexities such as illness or death and other matters that have greater consequences than what a young person is typically facing.

**What do you enjoy most about your job? What do you enjoy least about your job?**
I most enjoy the relationships I develop with my clients. A lot of my meetings take place around the dinner table. Clients share every aspect of their lives. This leads to a very close relationship that often lasts a lifetime.

I least like the stress that comes with managing a client's total financial nest egg. When the market drops, or two premier economists have completely opposing viewpoints, managing a portfolio is a big responsibility.

**Can you suggest a valuable "try this" for students considering a career in your profession?**
Pick a company, follow it on the stock market, watch it fluctuate and try to figure out why. I had one college course that really stuck with me and probably the only one that had any practical application to what I do. We bought a hypothetical stock portfolio at the beginning of the semester, explained why we liked the stock, and reported weekly on our transactions. Did we sell it and why? Did we keep it and why? We were graded on the rationale behind our moves.

Also, find a financial advisor you can interview or shadow. Offer to bring coffee or lunch. After all, you need to get used to asking for a meeting!

*This interview was originally published in 2016.*

# Police Officer/Detective

### Snapshot

**Career Cluster(s):** Government & Public Administration; Law, Public Safety, Corrections & Security

**Interests:** Law Enforcement; Investigation; Solving Crimes; Protecting the Public

**Earnings (Yearly Average):** $66,020

**Employment & Outlook:** As Fast As Average Growth Expected

## OVERVIEW

### Sphere of Work

Police officers protect lives and property. Detectives and criminal investigators, who are sometimes called agents or special agents, gather facts and collect evidence of crimes. With the increasing hold the internet and wireless networks have over our daily lives, it is no surprise that many police departments have units dedicated to the investigation of cyber related crimes, where some actions undertaken can include the identification of places (GPS tracking of suspects; pinpointing potential crime scenes) and persons (witnesses, suspects), email and IP tracing, etc.

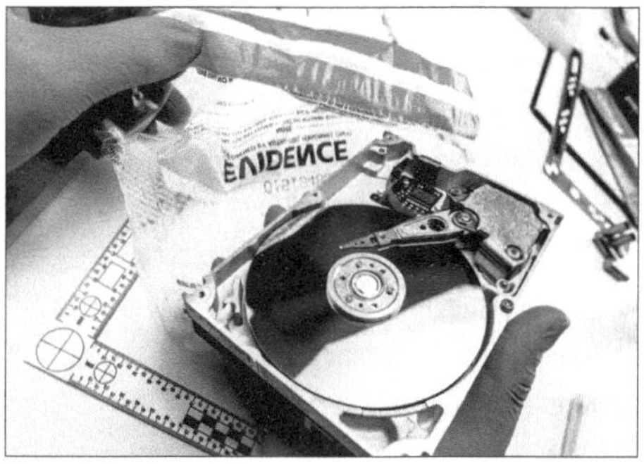

Photo via iStock.com/digicomphoto. [Used under license.]

### Work Environment

Police and detective work can be physically demanding, stressful, and dangerous. Officers must be alert and ready to react throughout their entire shift. Officers work at crime and accident scenes and encounter suffering and the results of violence. Cybercrime units in particular involve a good deal of monitoring of calls and emails from the public, who report incidents of potential fraud, unauthorized use of computers such as hacking and denial-of-service attacks, among other crimes.

### Profile

**Working Conditions:** Both Inside and Outside
**Physical Strength:** Heavy Work
**Education Needs:** High School Diploma; College Degree; Training Academy
**Licensure/Certification:** Not Required
**Opportunities for Experience:** On-the-Job Training
**Interest Score:** EIC

## Occupation Interest

Although a career in law enforcement may be stressful, many officers find it rewarding to help members of their communities. They enjoy the challenge of the job and the satisfaction of taking criminals off the streets. Further, detectives have a desire to piece together evidence to solve crimes, using the knowledge and tools at their disposal, as well as the input of forensic experts.

## A Day in the Life—Duties and Responsibilities

Job duties differ by employer and function, but police and detectives are required by law to write detailed reports and keep meticulous records. Most carry law enforcement equipment such as radios, handcuffs, and guns.

Detectives and criminal investigators are uniformed or plainclothes officers who gather facts and collect evidence related to criminal cases. They conduct interviews, examine records, monitor suspects, and participate in raids and arrests. Detectives typically investigate serious crimes, such as assaults, robberies, and homicides. In large police departments, detectives usually specialize in investigating one type of crime, such as homicide or fraud. They are typically assigned cases on a rotating basis and work on them until an arrest and trial are completed or until the case is dropped.

### Duties and Responsibilities

- Responding to emergency and nonemergency calls
- Patrolling assigned areas
- Conducting traffic stops and issuing citations
- Searching for vehicle records and warrants using computers in the field
- Obtaining warrants and arresting suspects
- Arresting people suspected of committing crimes
- Collecting and securing evidence from crime scenes
- Observing the activities of suspects
- Writing detailed reports and filling out forms
- Preparing cases and testifying in court

Federal Bureau of Investigation (FBI) agents, sometimes called special agents, investigate and pursue criminal cases that violate federal law. These agents are responsible for crimes against public agencies, such as Medicare fraud, or that cross state lines. In addition, federal agents may join or take over investigations of certain types of state cases, such as those related to prescription drugs or large sums of money.

Police and sheriff's patrol officers are the most common type of police and detectives, and they have general law enforcement duties. They wear uniforms that allow the public to easily recognize them as police officers. They have regular patrols and respond to emergency and nonemergency calls. During patrols, officers observe people and activities to ensure order and safety.

Some police officers work only on a specific type of crime, such as narcotics, or cybercrime. Officers, especially those working in large departments, may work in special units, such as mounted (horseback), motorcycle, or special weapons and tactics (SWAT). Typically, officers must work as patrol officers for a certain number of years before they are appointed to a special unit.

## OCCUPATION SPECIALTIES

### Fish and Game Warden

Fish and game wardens enforce fishing, hunting, and boating laws. They patrol fishing and hunting areas, conduct search and rescue operations, investigate complaints and accidents, and educate the public about laws pertaining to the outdoors. Federal fish and game wardens are often referred to as Federal Wildlife Officers.

### Transit and Railroad Police

Transit and railroad police patrol train yards and transportation hubs, such as subway stations. They protect property, employees, and passengers from crimes such as thefts and robberies. They remove trespassers from railroad and transit properties and check IDs of people who try to enter secure areas.

## WORK ENVIRONMENT

### Immediate Physical Environment

Police officers and detectives spend time both behind a desk and out in the field—this varies depending upon duty and seniority. They must be able to perform their duties in all weather, and under potentially life-threatening conditions.

Most police and detectives work full time. Paid overtime is common, and shift work is necessary to protect the public at all times. Special agents of the FBI must work at least 50 hours a week and are on call 24 hours a day, 7 days a week.

Some federal agencies, such as the FBI and U.S. Secret Service, require extensive travel, often on short notice. These agents may relocate a number of times over the course of their careers. Other agencies, such as U.S. Customs and Border Protection (CBP), may require work outdoors in rugged terrain and in all kinds of weather.

Police and sheriff's patrol officers and transit and railroad police have some of the highest rates of injuries and illnesses of all occupations. They may face physical injuries during conflicts with criminals and other high-risk situations.

### Human Environment

Police officers and detectives routinely interact with members of the public during their work, sometimes in highly stressful and dangerous situations. They must always remain focused and in control. Further, they rely on the support of their colleagues, especially during difficult situations. They may be called upon to testify in court and may also be interviewed by members of the media. Police officers are increasingly under public scrutiny and must be able to perform their duties without succumbing to distraction.

### Technological Environment

Police officers and detective utilize standard equipment such as flashlights, notepads, and smartphones, as well as firearms, electroshock weapons, handcuffs, and batons when subduing suspects. They are trained in the use of all equipment and must maintain their training. In the office, they may use database and email software to obtain and share information.

## EDUCATION AND TRAINING

### High School/Secondary

Police and detective applicants must have at least a high school diploma or equivalent. Interested students are encouraged to seek out a local Law Enforcement Exploring program, also known as "Police Explorers" programs, offered by Learning for Life—an affiliate of the Boy Scouts of America. These programs are accessible to young men and women who have completed the 6th grade and are younger than twenty-one years of age. The programs provide interested youth with hands-on experience in the criminal justice system, with training, practical activities, competitions, and more.

**Suggested High School Subjects**

- Algebra
- Biology
- Chemistry
- Civics
- Earth or Life or Physical Science
- Economics
- English
- Geometry
- History
- Math
- Physics or Forensic Science
- Sociology
- Statistics

**Related Career Pathways/Majors**

*Government & Public Administration Career Cluster*

- National Security Pathway

*Law, Public Safety, Corrections & Security Career Cluster*

- Law Enforcement Services Pathway

### *Transferable Skills and Abilities*

**Communication Skills**

- Speaking with people when gathering facts about a crime
- Expressing details about a given incident in writing

**Empathy**

- Understanding the perspectives of a wide variety of people
- Helping the public

**Good Judgment**

- Determining the best way to solve a wide array of problems quickly

**Leadership Skills**

- Motivating others and being comfortable as a highly visible member of the community
- Reacting quickly, responsibly, and calmly in emergency situations

**Perceptiveness**

- Anticipating a person's reactions and understanding why people act a certain way

**Physical Stamina**

- Passing required tests for entry into the field and keeping up with the daily rigors of the job

**Physical Strength**

- Exerting physical effort when apprehending suspects and assisting people in precarious situations

## Postsecondary

Police and detective applicants must have at least a high school diploma or equivalent, although some federal agencies and police departments may require that applicants have completed college coursework or a college degree. Many community colleges and 4-year colleges and universities offer programs in law enforcement and criminal justice. Knowledge of a foreign language is an asset in many federal agencies and geographical regions.

Fish and game wardens typically need a bachelor's degree; desirable fields of study include wildlife science, biology, or natural resources management. Federal Wildlife Officers and some state-level fish and game wardens typically do not need a bachelor's degree.

Federal agencies such as the FBI may require prospective detectives and investigators to have a bachelor's degree.

Many applicants for entry-level police jobs have completed some college coursework, and a significant number are college graduates.

**Related College Majors**

- Accounting
- Computer Science
- Criminal Justice
- Cyber Security
- Forensics
- Homeland Security
- Law
- Political Science
- Psychology
- Public Safety

## Adult Job Seekers

Because they need experience in law enforcement, detectives typically begin their careers as police officers. Special agent applicants for the FBI must have at least 2 years of full-time work experience, or 1 year of experience plus an advanced degree (master's or higher).

## Additional Requirements

Candidates for law enforcement appointment usually attend a training academy before becoming an officer. Training includes classroom instruction in state and local laws and constitutional law, civil rights, and police ethics. Recruits also receive training and supervised experience in subjects such as patrol, traffic control, firearm use, self-defense, first aid, and emergency response.

Federal law enforcement agents undergo extensive training, usually at the United States Marine Corps base in Quantico, Virginia, or at a Federal Law Enforcement Training Center (FLETC).

Some police departments have cadet programs for people interested in a career in law enforcement who do not yet meet age requirements for becoming an officer.

These cadets do clerical work and attend classes until they reach the minimum age requirement and can apply for a position with the regular force. Military or police experience may be considered beneficial for prospective cadets.

Cadet candidates must be U.S. citizens, usually be at least 18 years old, have a driver's license, and meet specific physical qualifications. Applicants may have to pass physical exams of vision, hearing, strength, and agility, as well as written exams. Candidates typically go through a series of interviews and may be asked to take polygraph (lie detector) and drug tests. A felony conviction may disqualify a candidate.

## EARNINGS AND ADVANCEMENT

Earnings of police officers and detectives depend upon the area of law enforcement and the employer—federal, state, or local. The median annual wage for police officers and detectives was $66,020 in 2021. The lowest 10 percent earned less than $40,420, and the highest 10 percent earned more than $105,540.

Many agencies provide officers with an allowance for uniforms, as well as extensive benefits and the option to retire at an age that is younger than the typical retirement age. Some police departments offer additional pay for bilingual officers or those with college degrees.

Police officers usually become eligible for promotion after a probationary period. Promotions to corporal, sergeant, lieutenant, and captain usually are made according to scores on a written examination and on-the-job performance. In large departments, an officer may be promoted to detective or to specialize in one type of police work, such as working with juveniles. Along with exam and performance scores, a bachelor's degree may be required for advancement to positions of lieutenant or higher rank.

## EMPLOYMENT AND OUTLOOK

Police and detectives held 808,200 jobs in 2021. Employment is expected to grow slower than average compared to all occupations through the year 2031, at a rate of 3 percent.

While a desire for public safety may result in a need for more officers, demand for employment is expected to vary depending upon location, driven largely by

**Fast Fact**

**Public complaints about internet crime are gathered by the FBI's Internet Crime Complaint Center (IC3). The Recovery Asset Team of the IC3 has helped victims of cybercrime by freezing hundreds of thousands of dollars using such complaints.**
Source: FBI.gov

Photo via iStock.com/domoyega. [Used under license.]

local and state budgets. Even when crime rates fall, demand for police services to maintain public safety is expected to continue.

About 68,500 openings for police and detectives are projected each year, on average, over the decade. Many of those openings are expected to result from the need to replace workers who transfer to different occupations or exit the labor force, such as to retire.

Job applicants may face competition because of relatively low rates of turnover. Applicants with a bachelor's degree and law enforcement or military experience, especially investigative experience, as well as those who speak more than one language, should have the best job opportunities.

Because the level of employment for police and detectives depends upon government spending, the number of job opportunities may vary from year to year and from place to place.

## Related Occupations

- Correctional Officer/Bailiff
- Emergency Management Director
- EMT/Paramedic
- Fire Inspector
- Firefighter
- Forensic Science Technician
- Private Detective/Investigator
- Probation Officer/Correctional Treatment Specialist
- Security Guard/Gambling Surveillance Officer

## MORE INFORMATION

**Bureau of Alcohol, Tobacco, Firearms and Explosives (ATF)**
99 New York Avenue NE
Washington, DC 20226
202.648.7080
www.atf.gov

**Exploring National Service Center**
1325 West Walnut Hill Lane
Irving, TX 75038
972.580.2433
exploring@lflmail.org
www.exploring.org
Offers the Law Enforcement Exploring program:
www.exploring.org/law-enforcement

**Federal Bureau of Investigation (FBI)**
935 Pennsylvania Avenue NW
Washington, DC 20535-0001
202.324.3000
www.fbi.gov

**National Association of Police Organizations (NAPO)**
317 South Patrick Street
Alexandria, VA 22314
703.549.0775
www.napo.org

**National Police Association (NPA)**
8710 Bash Street, Suite 501692
Indianapolis, IN 46250
302.469.1765
nationalpolice.org

**National Police Foundation (NPI)**
2550 S Clark Street, Suite 1130
Arlington, VA 22202
202.833.1460
info@policefoundation.org
www.policefoundation.org

**U.S. Customs and Border Protection (CBP)**
1300 Pennsylvania Avenue NW
Washington, DC 20229
202.325.8000
www.cbp.gov

**U.S. Department of Homeland Security (DHS)**
2707 Martin Luther King Jr. Avenue SE
Washington, DC 20528-0525
202.282.8000
www.dhs.gov

**U.S. Drug Enforcement Administration (DEA)**
600-700 Army Navy Drive
Arlington, VA 22202
202.307.1000
www.dea.gov

**U.S. Marshals Service (USMS)**
Arlington, VA
202.307.9100
www.usmarshals.gov

**U.S. Secret Service (USSS)**
245 Murray Lane SW, Building T-5
Washington, DC 20223
202.406.5708
www.secretservice.gov

---

*Stuart Paterson; updated by Daniel Vest*

# *Conversation With...*
# JAMAAL LITTLEJOHN

Fayetteville, North Carolina
Homicide detective, 5 years

**What was your individual career path in terms of education/training, entry-level job, or other significant opportunity?**

My father was active duty in the U.S. Army and my mother worked for Womack Army Medical Center at Fort Bragg, NC, as a licensed practical nurse. Growing up the only son of three children, I was my father's number one fan. I admired his work ethic and dedication to the military and our family. After completing only one semester at Fayetteville State University, I joined the Army Reserves in December 2007.

I attended basic training and AIT (Advanced Individual Training) in Fort Leonard Wood, Missouri, then entered the Military Police Corps in May 2008. During my first year as a reservist, I worked with a unit on Fort Bragg where we served garrison support for the military police units. During monthly drills and three weeks of annual training, I served as a military police officer.

In July 2009, I applied for and was accepted to the Fayetteville Police Academy in North Carolina. My wife and I had discussed the pros and cons of my becoming a police officer. A deciding factor was to put myself in a position to impact people of the community where I grew up. After graduating, I completed fourteen weeks of field training, which consisted of on-the-road training with an experienced office where I learned how to effectively enforce state law and interact with citizens.

I made a personal decision to work as a patrol officer for at least five years before transitioning to another unit within the department because I felt that would put me in a position to see and experience different incidents. It has always been my opinion that if I perfected my job at the lowest level, I would perform higher as I moved through the ranks.

In March 2015, I transferred to work as a detective in the aggravated assault unit and worked cases where citizens were victims of violent felony assaults. During this time, I used the communication skills I had perfected as a patrol officer to interview witnesses and suspects. After only five months in the aggravated assault unit, I transferred to the homicide unit, where I have worked for four years.

Numerous schools and training events helped me better serve the community while working death investigations. These include a general investigations course and PLI (Police Law Institute), as well as Vernon Geberth's Practical Homicide and Advanced Practical Homicide courses. Also, I have attended conferences as a member of the North Carolina Homicide Investigators Association.

I have been lead investigator on approximately seventeen homicide cases, and assisted on more than seventy-five. The unit has had only one unsolved homicide since December 2015.

**What are the most important skills and/or qualities for someone in your profession?**
Basic communication skills. On a daily basis, a homicide detective meets people on their worst days. You have to be able to show empathy and patience for the family because they just lost a loved one while also gathering pertinent information to solve the case for them. Depending on the circumstances surrounding a death, I have experienced family and friends withhold information to protect the deceased love one. It is important to make family, friends, and witnesses feel comfortable enough to give all the necessary information to solve the case as swiftly and efficiently as possible.

**What do you wish you had known going into this profession?**
How much one positive or negative contact with someone could change the way people view the entire profession. I was raised to treat everyone equally, and I carry that in my job. I am able to communicate with victims, witnesses, and suspects on the same level. My job as a patrol officer became harder after national incidents such as the shootings of Trayvon Martin and Michael Brown. Those and similar incidents changed people's opinions and made the way many people viewed law enforcement worse.

Growing up, I had very few positive experiences with local police officers. As a child, my older sister was approached by an officer as she walked with me from her friend's house. The officer asked my sister, who was sixteen at the time, "Do you want to get locked up?" I remember being angry; I was only five years old. As a teenager, I believed law enforcement officers could show up and conduct "regular business" because they felt like it and do this without any explanation to those they encountered. But, sitting and learning about why police officers did certain things was a life changer during my training to become one.

Now, even as a homicide detective, I spend the early part of some investigations building trust in order to move the case forward. That becomes easy with effective communication and sticking by my word. Along with other members of the homicide unit, I never make unrealistic promises to people. But, we do promise to put 100% into the investigation and provide answers to the families we serve.

**Are there many job opportunities in your profession? In what specific areas?**
One can work as a patrol officer or study to be promoted to detective where job opportunities include investigating property, robberies, or even juvenile and sex crimes. However, there are also numerous opportunities to work in law enforcement and investigate crimes on the federal level with the FBI, Secret Service, DEA, and others.

**How do you see your profession changing in the next five years? How will technology impact that change, and what skills will be required?**

I can only imagine how much technology could change my job. For instance, over the last four years in my unit, we have learned to identify suspects using phone records and social media. Those records also assisted detectives by tracking the current and previous locations of the victim and/or suspect.

**What do you enjoy most about your job? What do you enjoy least about your job?**

Mainly what I enjoy is solving cases and getting justice for the families affected. The real satisfaction, for me, comes in court when the suspect either is convicted or pleads guilty to the crimes committed.

I least enjoy the long hours we work. As seen on television, the first forty-eight hours of an investigation are the most important when it comes to finding a lead on the case. There is usually no pause or break for the first fifteen to twenty hours of a case. On multiple occasions, there have been times I have worked 20+ hours and gotten off work just to be called in for another investigation.

**Can you suggest a valuable "try this" for students considering a career in your profession?**

Volunteer at a local police department. Most have a ride along program where students can apply to ride with a patrol officer for a shift to see what the job is like on a daily basis. Internships within most departments are also available for college students.

*This interview was originally published in 2019.*

## *Conversation With...*
## JEFF LEDFORD

Chief of Police, Shelby, North Carolina
Law Enforcement, 26 years

**What was your individual career path in terms of education/training, entry-level job, or other significant opportunity?**

My first official training in law enforcement was in the police academy, specifically Basic Law Enforcement Training (BLET) at Gaston College in Dallas, NC. I chose law enforcement because I was drawn to the community service part of the job. I wanted to help people regardless of their issues.

After graduating, the Shelby Police Department hired me, and I began my career as all new recruits do, in the patrol division. I took several training classes to grow as an officer, such as radar operator, how to use breath alcohol testing instruments, and comparison and blood spatter analysis. Sometimes, veteran officers not only told me how to handle a situation better, but also showed me. I believe that much of the success I have had is directly attributed to those who helped me learn in my younger years.

As I progressed though my career, I realized that to excel I would need to pursue academics. I completed my bachelor's degree in criminal justice through University of Columbia and Gardner-Webb University.

I served as my agency's first full-time crime scene technician for several years in the patrol division, before a promotion to sergeant. Then, I served in administration as well as criminal investigations. Next, I became a lieutenant with the Problem Solving Unit, overseeing that and the Vice/Narcotics Unit. I was then promoted to the rank of captain as the Operations Commander, overseeing Shelby police's Patrol Division.

In 2004, my current chief retired and during the transition of preparing for a new chief, I was appointed as our agency's interim chief. This gave me an insight into the role's duties and helped me decide if that was a role I wanted someday. I knew I wanted to be chief but at the time, I felt too young and needed more experience.

I was selected Chief of Police for Shelby in 2008. I was honored. Every experience and training opportunity I had helped me to be successful in my current role.

**What are the most important skills and/or qualities for someone in your profession?**
One of the most important: humility. Focus on "who" you are more than "what" you are. While you lead the organization, you will only be able to lead people if they respect you. Realizing you are just like the other officers in your organization, but with a different set of responsibilities, will take you a long way.

Another important quality: relating to people. Leading an organization is predominately taking care of people and will help you more than anything as a leader. Being able to take care of people mindfully will help more than anything in a leadership role. Taking time every day to ensure the people you are responsible for are doing well and meeting all of their needs is important. It is fitting to say that if a person interacting with the public is not in a good place or has something bothering him/her, the service they deliver will not be the best it can be.

A final quality is being a cheerleader. While this sounds like an odd quality to have, it fits! Your job is to lead your organization and promote your agency in the best possible light to your staff and the community. You must always keep your staff upbeat and excited about their jobs because when they meet customers on the street, they will portray a positive image.

Also, keep the community excited about its police force. You must convey to the public that they are safe. Through good and bad times, they can rely on officers to take care of them, providing services they need.

**What do you wish you had known going into this profession?**
If people told me the things I would face when I was new to the profession, I would not have believed them. Some things you just have to experience for yourself. For example, if someone told me twenty years ago that law enforcement would have to work to keep up its public perception and acceptance, I would not have believed it.

However, I wish I had known how much the experienced people around would teach me. I was fortunate to have a lot of great people around me as I progressed through my career. The people who paved the way for all of us are a wealth of information. One lesson I learned was if you take care of your officers and staff inside the building, they will take care of the customer outside the building.

Also, I wish I had known the job's impact on my family. I have a wife, a son, and a daughter. My children are growing up in the era of social media, which means they have heard and read some really bad things about officers. Neither of them has ever really mentioned it to me, but I know it has to be hard on them. My wife deals with my job pulling me in different directions, but she also deals with walking me out every day knowing I may not return. It has to be difficult to be married to an officer. Families deserve as much credit as the officers.

Being chief, my family has to share me with the entire city of more than 20,000 people. Everyone expects you to be at his or her events and simultaneously, you have a family who expects you to make time for them, too. Time commitment is one of the biggest challenges for a chief and something everyone should discuss it with his or her family before accepting the position.

**Are there many job opportunities in your profession? In what specific areas?**
Turnover in law enforcement is becoming more of an issue. Agencies are looking for qualified people who have the drive and commitment to do the jobs. Finding people who are looking for a career helping others while understanding the dangers involved is becoming more difficult. Social media and the news often highlight the negative side of law enforcement, which could turn people away. Also, there are shows about law enforcement portraying it as a fun and low stress job. An applicant needs to look deeper into the profession before making that commitment.

The positions available include being a K9 officer, detective, vice agent to a traffic officer, or a state or federal government agent. State jobs include highway patrol, state investigator, or even North Carolina Wildlife Resources Commission. Federal jobs could range from being an FBI agent to a U.S. Marshal.

**How do you see your profession changing in the next five years? How will technology impact that change, and what skills will be required?**
We have seen more technology implemented over the past several years, a continuing trend. Many companies now offer products allowing us to do our jobs more efficiently and effectively through technology.

One of the biggest changes we have seen in our profession is social media, a great tool for law enforcement. We tell our story, connect with people, and inform people of an emergency or an event instantly. As an agency, we promote what we are doing and show people what we do that may not receive news coverage.

Social media can help us and be detrimental. Oftentimes, people will have limited facts or a tainted view of something and put it out. This can cause chaos and is something we have to be mindful of at all times. Officers now do their jobs every day knowing they are being filmed and reported on. Social media is here to stay, so we must change how we do business and learn to work with the medium to the agency's and public's benefit.

**What do you enjoy most about your job? What do you enjoy least about your job?**
Interacting with people in the department or the community and getting to know people and be part of their lives. Knowing I have impacted their lives in some small way is very rewarding, whether walking through a community and meeting people on their front porches or being around our headquarters and having a conversation with a member of our staff about his or her latest family vacation.

What I enjoy least is seeing tragedy and loss of life. Through the years, I have seen many young lives taken as a result of senseless violence. Nothing is harder than having to tell a parent their child is not coming home. No matter what the circumstance is, it never gets easier and each story is unique in its own way.

Tragedy and loss of life is not just confined to the community. In 2016, our agency was dealt the worst blow. We lost one of our own to violence—a K9 officer who served our city for more than 10 years. As a chief, it is the worst possible scenario you can face. While I always knew this could happen, I was not prepared to receive that call. To this day, we are dealing with the effects both mentally and as an agency. As chief, I realize each member of our organization is my responsibility. Losing one of them is a feeling of failure that cannot be described. It is something you have to learn to put into perspective and deal with every day while keeping your focus on taking care of the agency. It will change you, but only you can control in what way that change impacts you.

**Can you suggest a valuable "try this" for students considering a career in your profession?**

Research what the job truly entails. Do not consider TV show portrayals or the perceptions on the news or social media. Ask your local agencies if they have a ride-along program where you can ride and talk with an officer first-hand about what the job is like.

Secondly, remember all agencies conduct very thorough background investigations. Young people will make mistakes; it is a part of growing up. However, the decisions a young person makes can affect them for years to come. It is hard to deal with peer pressure, but knowing you are looking for a career in public service is a good incentive to make sure every decision you make takes you one step closer to that goal.

Lastly, find out what the job is about. Understand the risks involved, the lifestyle, and the impact it can have on your family. I truly believe law enforcement is the noblest profession in the world. To serve your community and protect those who cannot protect themselves is a feeling of accomplishment that cannot be described. It takes a special person to serve in law enforcement, and we will need people to continue to protect our cities, our states, and our country. We are always looking for the right people who are sometimes described as "the best of the best."

*This interview was originally published in 2019.*

## *Conversation With...*

## DAN ADAMS

Policeman, 18 years

**What was your individual career path in terms of education/training, entry-level job, or other significant opportunity?**

I was in the carpenters' union and getting laid off a lot. I had been in the Army for six years. I went over to Desert Storm. So I had the veterans' preference and was accepted into the Boston Police Academy. The physical training (P.T.) lasts the entire six months of the academy. You have to complete a 10-mile run. You have to complete a 1 1/2-mile run in an amount of time based on your age. You have to do a certain number of push-ups and a certain number of sit-ups in a two-minute period. You P.T. for 1 1/2 hours in the morning, then go into the classroom. You have to learn constitutional law and criminal law. It's tough.

I knew I wanted to get into the Drug Control Unit so I did an intense two-week course offered by the Drug Enforcement Agency (DEA). I was a patrolman in the drug unit for 13 years, and recently became a detective. You take a written exam and if you get a passing grade, you move onto oral questioning before a board. I have my Master's and Bachelor's in criminal justice.

**What are the most important skills and/or qualities for someone in your profession?**

Having a good memory is really up there, which not everybody has. You get thrown a lot of information at roll call. There's the Boston Regional Intelligence Center (BRIC) reports. You have to remember details like the make of a car, or remember people. Secondly, you definitely have to have people skills. No one likes being arrested. If you explain what's happening and talk them through the process, you're going to get a lot less resistance.

Some patrolmen have a bad attitude about calls like a bicycle theft. But that person probably paid $2,000 for that bike and they're really bummed out. That person may have never called 911 in their life, and when they do it's not a good experience. You have to be professional on every call. Some people sort of lose their empathy as the years go on.

**What do you wish you had known going into this profession?**

Nothing, really. Now that I'm close to being a sergeant, I wish I had studied more so I could do well on that exam.

**Are there many job opportunities in your profession? In what specific areas?**

With mandatory retirement, there are always people retiring. Boston has 2,200 officers. We have a dive team. If you're into dogs, you can get on the canine unit. We have a bomb unit. I'm in the Drug Control Unit. We have community relations, a bank robbery task force, crimes against children...I could go on and on. You can do so many things in law enforcement. It's great.

**How do you see your profession changing in the next five years? What role will technology play in those changes, and what skills will be required?**

We're constantly getting upgrades and training. Finding people is so much easier; there's just so much of a paper trail. Just Google yourself-you'll be amazed. And the police have different search engines. In the old days, you would go out to location, asking, "Hey, have you seen Johnny?" Today, there are 81 cameras around the city. Within five years, that's going to be more like 800 cameras. I can be sitting at my desk manipulating a camera and watching.

**What do you enjoy most about your job? What do you enjoy least?**

We call it "hunting." It's going out and catching the bad guys. You might watch a kid in the freezing cold. For 45 minutes, he's standing around, he's checking his cell phone, he's watching the cars go by. Then a car pulls up and he goes around the block for one minute. It's what we call "a meaningless ride." Right now I'm working on a really big search warrant that we're going to execute tomorrow. I love that.

Because it's Civil Service—and I shouldn't complain about that, because I benefited from it-but it's so, so hard to get rid of the dead weight. People you would never want to work with. And then you have the people who are at the other end. A little power makes them goofy. Of course, there are plenty of really great guys that you love to work with, who have a really great work ethic and who are doing a lot to make Boston a better city.

**Can you suggest a valuable "try this" for students considering a career in your profession?**

Try to get a seasonal job in a tourist area that hires officers for the summer. People go for security guard jobs. An EMT position would also be good training.

*This interview was originally published in 2014.*

# Private Detective/Investigator

**Snapshot**

**Career Cluster(s):** Law, Public Safety, Corrections & Security

**Interests:** Law & Justice; Investigation; Surveillance

**Earnings (Yearly Average):** $59,380

**Employment & Outlook:** Fast As Average Growth Expected

## OVERVIEW

### Sphere of Work

Private detectives and investigators search for information about legal, financial, and personal matters. They offer many services, such as verifying people's backgrounds and statements, finding missing persons, and investigating computer crimes. In investigating cybercrimes, private detectives gather and recover digital information regarding a variety of case types, such as family law cases, infidelity claims, domestic disputes, employment disputes, personal injury, and more.

### Work Environment

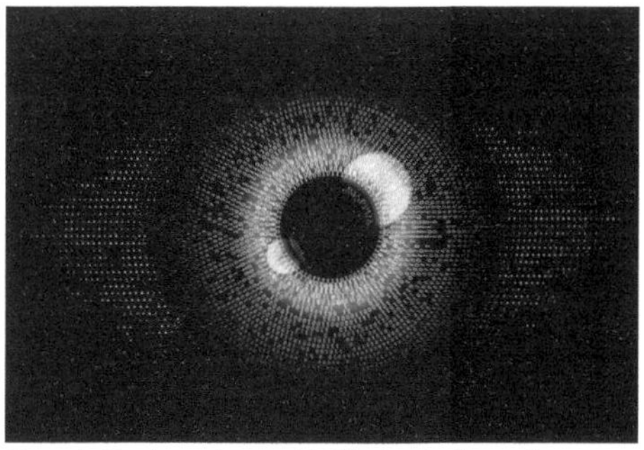

Photo via iStock.com/seamartini. [Used under license.]

The work environment for private detectives and investigators ranges from office work, such as computer research and phone calls, to field work such as interviewing and surveillance. Private detectives and investigators often work irregular hours because they conduct surveillance and contact people outside of normal work hours. They may work early mornings, evenings, weekends, and holidays.

## Profile

**Working Conditions:** Both Inside and Outside
**Physical Strength:** Medium Work
**Education Needs:** High School Diploma; Work Experience
**Licensure/Certification:** Varies
**Opportunities for Experience:** On-the-Job Training
**Interest Score:** EC

### Occupation Interest

Private detectives and investigators are inquisitive, highly analytical individuals who enjoy taking on challenging cases and solving real-world mysteries, at the behest of clients. They typically come from related fields such as law enforcement, the military, or intelligence, where the required skill set is honed through years of work experience.

### A Day in the Life—Duties and Responsibilities

Private detectives and investigators offer many services for individuals, attorneys, and businesses. Examples include performing background checks, investigating employees for possible theft from a company, proving or disproving infidelity in a divorce case, and helping to locate a missing person.

Private detectives and investigators use a variety of tools when researching the facts in a case. Much of their work is done with a computer, allowing them to obtain information such as telephone numbers, details about social networks, descriptions of online activities, and records of a person's prior arrests. They make phone calls to verify facts and interview people when conducting a background investigation.

Detectives also conduct surveillance when investigating a case. They may watch locations, such as a person's home or office, often from a hidden position. Using cameras and binoculars, detectives gather information on people of interest.

## Duties and Responsibilities

- Interviewing people to gather information
- Searching online, public, and court records to uncover clues
- Conducting surveillance
- Collecting evidence for clients
- Checking for civil judgments and criminal history

Detectives and investigators must be mindful of the law when conducting investigations. Because they lack police authority, their work must be done with the same authority as a private citizen. As a result, detectives and investigators must have a good understanding of federal, state, and local laws, such as privacy laws, and other legal issues affecting their work. Otherwise, evidence they collect may not be usable in court, and they could face prosecution.

## OCCUPATION SPECIALTIES

**Skip Tracer**
Skip tracers specialize in locating people whose whereabouts are unknown. For example, debt collectors may employ them to locate people who have unpaid bills.

## WORK ENVIRONMENT

### Immediate Physical Environment

Private detectives and investigators work in many environments, depending on the case. Some spend more time in offices, researching cases on computers and making phone calls. Others spend more time in the field, conducting interviews or performing surveillance. In addition, private detectives and investigators may have to work outdoors or from a vehicle, in all kinds of weather, to obtain the information their client needs.

### Human Environment

Although investigators often work alone, some work with others while conducting surveillance or executing large, complicated assignments. Some may also choose to employ a team of investigators. Investigators must be comfortable dealing with clients and members of the public, sometimes in highly stressful situations. They may also need to deal professionally with the police and give evidence in court.

### Technological Environment

Private detectives and investigators must be able to use all tools at their disposal for solving cases, including standard computer software, databases, and a familiarity with social media and the internet. Sophisticated camera equipment may also be required, as well as standard field tools such as binoculars. Private detectives and investigators should ideally have their own mode of transportation because travel is often required.

## EDUCATION AND TRAINING

### High School/Secondary

Education requirements vary greatly with the job, but most jobs require a high school diploma. Students with an interest in this line of work should prepare as though for a career in law enforcement, with courses such as English, civics, history, psychology, and sociology.

**Suggested High School Subjects**

- Algebra
- Biology
- Chemistry
- Civics
- Earth or Life or Physical Science
- Economics
- English
- Forensic Science
- Geometry
- History
- Physics
- Psychology
- Sociology
- Statistics

**Related Career Pathways/Majors**

*Law, Public Safety, Corrections & Security Career Cluster*

- Law Enforcement Services Pathway
- Security & Protective Services Pathway

#### *Transferable Skills and Abilities*

**Communication Skills**

- Listening carefully and asking appropriate questions when interviewing a person of interest

**Decision-making Skills**

- Thinking on one's feet and making quick decisions, based on the limited information that they have at a given time

**Inquisitiveness**

- Wanting to ask questions and search for the truth

**Patience**

- Spending long periods conducting surveillance while waiting for an event to occur
- Investigating cases may take a long time, and they may not provide a resolution quickly—or at all

**Resourcefulness**

- Working persistently with whatever leads are available, no matter how limited, to determine the next step toward their goal
- Anticipating what a person of interest will do next

### Postsecondary

Private detectives and investigators typically need several years of work experience and a high school diploma. In addition, the vast majority of states require private detectives and investigators to have a license.

Education requirements vary greatly with the job, but most jobs require a high school diploma. Some jobs may require a 2- or 4-year degree in a field such as criminal justice.

Private detectives and investigators must typically have previous work experience, usually in law enforcement, the military, or federal intelligence, each with its own educational requirements. Those in such related jobs, who retire after 20 or 25 years of service, may become private detectives or investigators in a second career.

Other private detectives and investigators may have previously worked as bill and account collectors, claims adjusters, paralegals, or process servers.

**Related College Majors**

- Criminal Justice/Police Science
- Cultural/Archaeological Resources Protection
- Cyber/Computer Forensics and Counterterrorism
- Financial Forensics & Fraud Investigation
- Law Enforcement Recordkeeping and Evidence Management
- Maritime Law Enforcement
- Protective Services Operations

## Professional Certification and Licensure

Most states require private detectives and investigators to have a license. Check with your state for more information; Professional Investigator Magazine has links to most states' licensing requirements. Because laws often change, jobseekers should verify the licensing laws related to private investigators with the state and locality in which they want to work.

Candidates may also obtain certification, although it is not required for employment. Still, becoming certified through professional organizations can demonstrate competence and may help candidates advance in their careers.

For investigators who specialize in negligence or criminal defense investigation, the National Association of Legal Investigators (NALI) offers the Certified Legal Investigator (CLI) certification. For other investigators, the American Society for Industrial Security (ASIS) International offers the Professional Certified Investigator certification.

## Additional Requirements

Most private detectives and investigators learn through on-the-job training, typically lasting between several months and a year.

**Fast Fact**

**Part of the work cyber investigators do includes collecting social media posts, which can then be used in court as additional evidence to demonstrate foundation and character, support alibis, and provide important details pertinent to court proceedings.** Source: boscolegal.org

Photo via iStock.com/cofotoisme. [Used under license.]

Although new investigators must learn how to gather information, additional training depends on the type of firm that hires them. For example, investigators may learn to conduct remote surveillance, reconstruct accident scenes, or investigate insurance fraud. Corporate investigators hired by large companies may receive formal training in business practices, management structure, and various finance-related topics.

## EARNINGS AND ADVANCEMENT

Earnings depend on the experience of the investigator, the nature of the cases they work, and the clients for whom they work. Median annual earnings of private detectives and investigators were $59,380 in 2021. The lowest 10 percent earned less than $32,130, and the highest 10 percent earned more than $98,070.

## EMPLOYMENT AND OUTLOOK

Private detectives and investigators held 37,000 jobs in 2021. Eight percent were self-employed. Employment is expected to grow much faster than all occupations through the year 2031, at a rate of 6 percent. However, because it is a small occupation, the fast growth will result in only about 2,100 new jobs over the 10-year period.

Continued lawsuits, fraud and other crimes (including an increasing prevalence of cybercrimes), and interpersonal mistrust create demand for investigative services in industries such as legal services.

Background checks will continue to be a source of work for some investigators, as online investigations are not always sufficient.

Strong competition for jobs can be expected because private detective and investigator careers attract many qualified people, including relatively young retirees from law enforcement and the military.

Candidates with related work experience, as well as those with strong interviewing skills and familiarity with computers, may find more job opportunities than others.

### Related Occupations

- Accountant/Auditor
- Bill/Account Collector
- Claims Adjuster/Appraiser/Examiner/Investigator
- Financial Examiner
- Fire Inspector
- Forensic Science Technician
- Paralegal/Legal Assistant
- Police Officer/Detective
- Security Guard/Gambling Surveillance Officer

## MORE INFORMATION

**ASIS International**
1625 Prince Street
Alexandria, VA 22314-2882
703.519.6200
asis@asisonline.org

**Council of International Investigators (CII)**
P.O. Box 1015
St. Helena, SC 29920
917.292.0613
lec@cii2.org
www.cii2.org

**International Intelligence Network (INTELLENET)**
P.O. Box 350
Gladwyne, PA 19035
734.320.9240
Intellenet@intellenet.org
www.intellenet.org

**National Association of Legal Investigators (NALI)**
P.O. Box 278
Bath, MI 48808
517.702.9835
www.nalionline.org

**National Association of Professional Process Servers (NAPPS)**
P.O. Box 4547
Portland, OR 97208-4547
503.222.4180
napps.org

**National Council of Investigation and Security Services (NCISS)**
P.O. Box 200615
Evans, CO 80620
800.445.8408
www.nciss.org

**National Defender Investigator Association (NDIA)**
P.O. Box 1900
San Marcos, CA 92079-1900
760.290.3909
nationaldefender@gmail.com
www.ndia.net

**Professional Background Screening Association (PBSA)**
110 Horizon Drive, Suite 210
Raleigh, NC 27615
919.459.2082
info@thepbsa.org
thepbsa.org

**World Association of Detectives (WAD)**
P.O. Box 20030
Evans, CO 80620
443.982.4586
Wad@Wad.net
www.wad.net

---

*Stuart Paterson; updated by Daniel Vest*

# *Conversation With...*
# PAMELA BEASON
Private Investigator, 6 years

**What was your individual career path in terms of education/training, entry-level job, or other significant opportunity?**

Before becoming a professional investigator, I worked as a technical writer/ editor, mechanical/architectural drafter, college instructor and geological research technologist. I got two college degrees before completing a certification program in Investigation from the University of Washington, and then studying for the state exam for individual investigator. A few years later I met my business partner, Molly Monahan, and we both studied for the state agency exam and, after qualifying, set up our partnership at Sirius Investigations.

I also attended the Redmond Citizens Police and Fire Academy to learn the language and methodology employed by first responders, and I continue to attend educational seminars in specialized legal matters. My interests are pre-trial investigations and product liability issues.

**What are the most important skills and/or qualities for someone in your profession?**

There are very few employee jobs for investigators; most of us are independent contractors, so the ability to run your own business is paramount. As for the investigation work itself, you need to be able to think analytically without preconceived notions. You must be able to systematically sort through reams of data to find the important bits. You have to be able to communicate clearly in person and in writing; organize your time efficiently; and persuade all kinds of people to give you information. And every private investigator (PI) must have discretion, of course. It's our legal obligation to keep our clients' secrets.

**What do you wish you had known going into this profession?**

I wish I had known that most jobs would be only a few hours long, thus necessitating a lot of unpaid hours speaking to potential clients, as well as creating a major paperwork and bookkeeping hassle.

**Are there many job opportunities in your profession? In what specific areas?**

There are many more opportunities in large urban areas than in towns with smaller populations. Generally speaking, no matter where you live, you must set up and run your own business, because there are very few investigation agencies with employees. In some cities, public defender offices rely on private investigators for defense investigations, so that can be a big source of jobs, although they are usually not the best-paying contracts. There is always a need for surveillance—usually these jobs are to investigate insurance fraud (such as faking a disability) and document child custody issues (proving who is actually caring for the child and who is living in the home). Another fast-growing segment of investigative work involves computer analysis—recovering data from hard drives, tracking emails, and so forth. If you live in a large urban area, you can probably specialize. If you live in a smaller town such as Molly and I do, you have to handle all sorts of cases to make a living.

**How do you see your profession changing in the next five years? What role will technology play in those changes, and what skills will be required?**

I don't think the profession will change too much—there will always be a need for investigators to interview people, gather evidence, and file reports or testify in court. We already have to continually learn to use all sorts of new databases and interfaces to get information and we must adapt to all sorts of technology—computers, cameras, recorders, smart phones, etc. This need to constantly learn on the job will continue.

**What do you enjoy most about your job? What do you enjoy least?**

Because I'm a writer, too, I most enjoy interviewing witnesses and participants for upcoming court cases. I don't like surveillance work; it's very tedious, and my business partner (and almost every investigator I know) agrees with that. What Molly enjoys most is the joy of locating a "lost" loved one for a client, and the gratitude of clients when we get the results they want.

**Can you suggest a valuable "try this" for students considering a career in your profession?**

To test your interview skills, I'd suggest you interview someone and take notes (PIs often are not allowed to use recording devices), then write up that interview and share it with the interviewee to see how well you do. To test surveillance skills, spy on someone you know (warn them that you might do this so you won't scare them if they see you, but don't tell them when) and take time-stamped photos or video, then transform that into a package you could hand or email to a client.

I get asked so often what it's like to be an investigator that I wrote a little ebook to explain the job and the skills. It's called "So You Want to Be a PI?" by Pamela Beason.

*This interview was originally conducted in 2014.*

# Software Developer/ Quality Assurance Analyst/Tester

### Snapshot

**Career Cluster(s):** Information Technology; Manufacturing; Science, Technology, Engineering & Mathematics

**Interests:** Software; Computers; Product Testing; Analysis

**Earnings (Yearly Average):** $109,020

**Employment & Outlook:** Much Faster Than Average Growth Expected

## OVERVIEW

### Sphere of Work

Software developers create the computer applications that allow users to do specific tasks and the underlying systems that run the devices or control networks. Software quality assurance analysts and testers design and execute software tests to identify problems and learn how the software works. One major type of software being designed and tested is cybersecurity software, which aims to protect individual consumers and business clients from attacks on the web.

Photo via iStock.com/Laurence Dutton. [Used under license.]

## Work Environment

Software development takes place in office settings, and is a collaborative process with teams of people performing various duties to create the finished whole.

## Profile

**Working Conditions:** Inside Work
**Physical Strength:** Medium Work
**Education Needs:** Bachelor's Degree; Master's Degree
**Licensure/Certification:** Not Required
**Opportunities for Experience:** Coursework; Internship
**Interest Score:** ICR

## Occupation Interest

Software developers and quality assurance analysts and testers all have strong foundations in computer programming and coding, and apply that knowledge and interest in creating new software programs in different aspects of the process.

## Duties and Responsibilities

**Software Developers:**

- Analyzing users' needs and then designing and developing software to meet those needs
- Recommending software upgrades for customers' existing programs and systems
- Designing each piece of an application or system and planning how the pieces will work together
- Creating a variety of models and diagrams showing programmers the software code needed for an application
- Ensuring that a program continues to function normally through software maintenance and testing
- Documenting every aspect of an application or system as a reference for future maintenance and upgrades

**Software Quality Assurance Analysts and Testers:**

- Creating test plans, scenarios, and procedures for new software
- Identifying project risks and recommending steps to minimize those risks
- Implementing software testing, using either manual or automated programs and exploratory testing, and evaluate results
- Documenting and reporting defects or problems with software
- Providing feedback to software developers and stakeholders regarding usability and functionality

## A Day in the Life—Duties and Responsibilities

Software developers, quality assurance analysts, and testers are involved in the entire process of creating a software program. Developers may begin by asking how the customer plans to use the software so that they can identify the core functionality the user needs. Software developers also determine other requirements, such as security. They design the program and then work closely with programmers, who write computer code. However, some developers write code themselves instead of giving instructions to programmers.

Software quality assurance analysts and testers design and execute systems to check the software for problems. As part of their testing, these workers document and track the software's potential defects or risks. They also assess its usability and functionality to identify difficulties a user might have. After completing testing, they report the results to software or web developers and review ways to solve any problems they found.

After the program is released to the customer, a developer may perform upgrades and maintenance. Quality assurance analysts and testers run manual and automated checks to look for errors and usability problems once the software is released and after any upgrades or maintenance.

Developers who supervise a software project from the planning stages through implementation sometimes are called information technology (IT) project managers. These workers monitor the project's progress to ensure that it meets deadlines, standards, and cost targets.

## OCCUPATION SPECIALTIES

**Applications Software Developers**
Applications software developers design computer applications, such as games, for consumers. They may create custom software for a specific customer or commercial software to be sold to the general public. Some applications software developers create databases or programs for use internally or online.

**Software Engineers**
Software engineers take a broad view of a project's system and software requirements, planning its scope and order of work. These workers may direct software developers, quality assurance analysts, and testers.

**Systems Software Developers**
Systems software developers create the operating systems for the public or specifically for an organization. These operating systems keep computers functioning and control most of the consumer electronics in use today, including those in cell phones and cars, and also including AI capabilities. Often, systems software developers also build the interface that allows users to interact with the computer.

## WORK ENVIRONMENT

### Immediate Physical Environment

Developing software is usually a collaborative process. As a result, developers, quality assurance analysts, and testers work on teams with others who also contribute to designing, developing, and programming successful software.

Most software developers, quality assurance analysts, and testers work full time.

### Human Environment

Software developers and quality assurance analysts and testers must be able to maintain their focus and work with others for long periods during the development phase, and be meticulous about their testing during the end-phase and routine maintenance. Software is becoming ever more complex, including the integration of artificial intelligence (AI) components, and as these systems become more commonplace in everyday life, their safe operation is paramount. Developers and analysts must take direction from their clients, and may report to more senior developers.

### Technological Environment

Software developers and quality assurance analysts and testers must be apprised of the latest developments in computer programming and software engineering in order to keep pace with trends, and the desires of their clients. This includes familiarity with AI capabilities and other advanced systems.

## EDUCATION AND TRAINING

### High School/Secondary

High school students interested in becoming software developers or quality assurance analysts or testers should concentrate on computer science, math, and other science-related courses, as well as participating in after-school computer activities to hone their skills. They should plan on attending college or university to further their knowledge of programming and software development.

**Suggested High School Subjects**

- Algebra
- Applied Physics
- Biology
- Calculus
- Chemistry
- Computer Programming
- Computer Science
- Earth or Life or Physical Science
- Economics
- English
- Geometry
- History

- Political Science
- Pre-calculus
- Trigonometry

**Related Career Pathways/Majors**

*Information Technology Career Cluster*

- Information Support & Services Pathway
- Programming & Software Development Pathway

*Manufacturing Career Cluster*

- Quality Assurance Pathway

*Science, Technology, Engineering & Mathematics Career Cluster*

- Engineering & Technology Career Pathway

## Transferable Skills and Abilities

**Analytical Skills**

- Evaluating users' needs and then designing software to function properly and meet those needs

**Communication Skills**

- Giving clear instructions and explaining problems that arise to other team members involved in development
- Explaining to nontechnical users, such as customers, how the software works and answering any questions that arise

**Creativity**

- Being innovative in their approaches to designing, identifying problems with, and improving computer software

**Detail-oriented**

- Concentrating on many parts of an application or system at the same time, and paying attention to detail when looking for potential areas of user error

**Interpersonal Skills**

- Working well with others who contribute to designing, programming, and testing successful software

**Problem-solving Skills**

- Solving problems that arise throughout the design process

## Postsecondary

Software developers, quality assurance analysts, and testers typically need a bachelor's degree in computer and information technology or a related field, such as engineering or mathematics. Computer and information technology degree programs cover a broad range of topics. Students may gain experience in software development by completing an internship, such as at a software company, while in college. For some software developer positions, employers may prefer that applicants have a master's degree.

Although writing code is not their primary responsibility, developers must have a strong background in computer programming. They usually gain this experience in school. Throughout their career, developers must keep up to date on new tools and computer languages.

**Related College Majors**

- Business Administration
- Computer Science
- Data Science
- Entrepreneurship
- Information Technology

- Mathematics
- Software Development
- Software Engineering

### Adult Job Seekers

Adults from related fields such as software engineering may be able to transition into development or analysis, however extensive coursework and practical experience may be required in order to gain familiarity with the coding and programming needed to perform the duties in this area.

## EARNINGS AND ADVANCEMENT

Earnings of software developers, quality assurance analysts, and testers depend on their education and experience, employer, and the type of software being developed and tested. The median annual wage for software developers was $120,730 in 2021. The lowest 10 percent earned less than $64,470, and the highest 10 percent earned more than $168,570.

The median annual wage for software quality assurance analysts and testers was $98,220 in 2021. The lowest 10 percent earned less than $48,960, and the highest 10 percent earned more than $153,250.

Software developers, quality assurance analysts, and testers may receive benefits such as paid vacations, holidays, and sick days; life and health insurance; and retirement benefits, usually paid by the employer.

Software developers can advance to become project management specialists or computer and information systems managers, positions in which they oversee the software development process.

## EMPLOYMENT AND OUTLOOK

There were approximately 1,622,200 software developers, quality assurance analysts, and testers employed nationally in 2021. Employment is projected to grow 25 percent from 2021 to 2031, much faster than the average for all occupations.

The need for new applications on smart phones and tablets will help increase the demand for software developers.

## Famous First

**Commercial anti-virus software first appeared in 1987, when German developers Andreas Lüning and Kai Figge innovated a program for the Atari ST. Meanwhile, in the U.S., John McAfee (pictured) founded McAfee and released VirusScan.**
Source: cybermagazine.com

Photo by Gage Skidmore, via Wikimedia Commons.

The health and medical insurance and reinsurance carriers industry will need innovative software to manage new healthcare policy enrollments and administer existing policies digitally. As the number of people who use this digital platform increases over time, demand for software developers, quality assurance analysts, and testers will grow.

Software developers, quality assurance analysts, and testers are likely to see new opportunities because of an increase in the number of products that use software. For example, computer systems are routinely built into consumer electronics and other products, such as cell phones and appliances. Demand for more advanced systems such as AI are also expected to increase, requiring programmers and analysts with special skillsets to design these platforms.

Concerns over threats to computer security could result in more investment in security software to protect computer networks and electronic infrastructure.

### Related Occupations

- Computer/Information Research Scientist
- Computer/Information Systems Manager
- Computer Hardware Engineer
- Computer Network Architect
- Computer Programmer
- Computer Support Specialist
- Computer Systems Analyst
- Database Administrator/Architect
- Information Security Analyst
- Mathematician/Statistician
- Postsecondary Teacher
- Web Developer/Digital Designer

## MORE INFORMATION

**American Society for Quality (ASQ)**
P.O. Box 3005
Milwaukee, WI 53201-3005
800.248.1946
asq.org

**Association for Computing Machinery (ACM)**
1601 Broadway, 10th Floor
New York, NY 10019-7440
212.869.7440
www.acm.org

**Association for Software Testing (AST)**
marketing@associationforsoftwaretesting.org
associationforsoftwaretesting.org

**CompTIA**
3500 Lacey Road, Suite 100
Downers Grove, IL 60515
866.835.8020
membership@comptia.org
www.comptia.org

**Computing Research Association (CRA)**
1828 L Street NW, Suite 800
Washington, DC 20036-4632
202.234.2111
cra.org

**IEEE Computer Society**
2001 L Street NW, 700
Washington, DC 20036-4928
202.371.0101
help@computer.org
www.computer.org

**National Center for Women & Information Technology (NCWIT)**
1909 26th Street, 2nd Floor
Boulder, CO 80302
303.735.6671
info@ncwit.org
www.ncwit.org

**Quality Assurance Association (QAA)**
1301 Riverplace Boulevard, Suite 800
Jacksonville, FL 32207
QAA@qualityassuranceassociation.com
www.qualityassuranceassociation.com

**Society of Quality Assurance (SQA)**
154 Hansen Road, Suite 201
Charlottesville, VA 22911
434.297.4772
sqa@sqa.org
www.sqa.org

---

*Stuart Paterson; updated by Daniel Vest*

## *Conversation With...*

## BRIAN THARP

Principal and co-founder, ZENxd
HQ Santa Cruz, CA
Technology application design, 23 years

**What was your individual career path in terms of education/training, entry-level job, or other significant opportunity?**

I'm a principal and co-founder of a technology application design firm, but I didn't start in this field. I was a Fine Arts major at Arizona State University with an emphasis on printmaking. In the middle of my junior year, I left college. I was getting burned out because while a student I was working full-time to pay for school. I'd planned to take a break and return to college after earning more money to pay for it, but I never did. I realized there wasn't much opportunity in printmaking and instead I got a job designing circuit boards. I had a generous boss who taught me how to do the work. It was interesting, creative work that challenged me. Then I got a job testing software and networking equipment and learned about packet transfer and things I never expected to learn. I ended up setting up the company's intranet, and that's how I was introduced to web technology and design. As the industry evolved, so did a focus on human-centered design. That was my bridge back to my creative self. Before starting my company, I'd worked for more than a half dozen companies, designing systems that help people do things like comparison shop, or help businesses deploy campaigns faster, at lower costs and with higher interaction with their targeted audience.

In 2008, a colleague, Bill Daggett, and I partnered and founded ZENxd, a boutique UX Design firm. We are a distributed team, working from multiple locations, including Portland, Oregon, where I'm based, and Santa Cruz, California, where Bill is [based]. Bill and I had worked for a company called SideStep, Inc. where we designed the first application that allowed travelers to use a single website to search multiple sites for information on flights, lodging, car rental, guides, and deals. This was integrated into KAYAK's platform when KAYAK bought SideStep.

ZENxd designs applications for websites, mobile or any other type of device where there is human interaction required. We research users' needs, what they are looking for, how to engage them, and how long they interact with the applications. We focus on the human side, the users, and define strategies that help people solve a problem or reach a goal. Typically, a UX strategy defines several "digital touch points" that are needed to create a holistic experience. We just did a big project with Shutterfly. We helped redesign their manufacturing application by unifying it into a single platform so it's easier for their workers to learn and operate. We work on projects that could be something as big as a whole production platform, to as small as a smart watch application.

I've always loved tackling big design problems. When you work for a company, the meaty problems get solved, and you are left to "maintain" the experience. I would get the itch to jump ship and find a new problem to solve. So, I co-founded ZENxd to ensure I'd always have new challenges and get to wear a lot of hats.

**What are the most important skills and/or qualities for someone in your profession?**
Understanding the fundamental principles of design is key to knowing how a design is perceived, how people will learn from it, how to make it more appealing and useful.

You don't need to learn everything; it's a matter of starting with your strengths. If you understand good design principles—visual, color—start with those.

A good application design also involves research. If you're inquisitive and a people person, you may want to focus on research. Start there and begin to broaden.

Empathy, active listening, observation, and curiosity are important traits. Think about the lived experience of someone who will use the application and how they perceive or understand the design.

You have to collaborate with people, facilitate and bring in different views. It's critical to communicate clearly with people who perform different functions of work on the project.

**What do you wish you had known going into this profession?**
How nebulous the profession would prove to be. Even today it's hard for people to understand what I do. People say "you make it look good" or "you make it look simple," and there's more to it. Making something look simple is extremely hard. Making design decisions is a challenge because you have lots of needs and resources that are in conflict.

**Are there many job opportunities in your profession? In what specific areas?**
There are a ton of opportunities, and they're becoming more specialized and diverse due to advances in technology and the proliferation of disparate devices. There are jobs that focus on user research, experience design (UX), visual design (UI), service design, design strategy, design leadership, etc. Areas to specialize in include mobile applications, voice enabled devices, augmented reality, artificial intelligence.

**How do you see your profession changing in the next 5years? How will technology impact that change, and what skills will be required?**
It's harder for companies to differentiate on technology. Today's consumers have grown up in a digital world. Technology has changed expectations. If a customer has a bad experience, another option is just a tap away. To maintain market share, companies have shifted their focus to customer engagement. As a result, Experience Design has become a critical component to achieving long-term success.

Voice assistance, machine learning, and artificial intelligence are becoming more predictive. We are focused on developing relationships, learning about the customer, and anticipating the customer's needs.

There's growing emphasis on healthy experiences—understanding an application's impact on the individual's well-being. How well did the experience promote healthy habits, or does it inhibit healthy habits? Thinking about well-being is a big part of what we do because we're designing for humans.

There's an overdue emphasis on design justice and inclusive design. We are hyper-focused on improving our methods and practices to ensure our designs produce quality experiences for as many people as possible. If you neglect a segment of the population, you miss reaching potential customers. You also risk damaging your brand if you lack the care to ensure your products are usable and accessible.

**What do you enjoy most about your job? What do you enjoy least about your job?**
I get to speak with and learn about people from all walks of life. It's fun, enlightening, and even emotional at times. It's my job to improve their lives, and that enriches me.

I like learning about a variety of topics. We've designed experiences in health care, travel, legal, production accounting—all kinds of domains. You're really trying to understand the industries and it's never boring.

What I enjoy least is feeling in a constant state of critique. As a designer, everything you produce is subject to evaluation and criticism. You must be prepared to provide a rationale for your designs. While critique can be hard, it continually presents opportunities for growth and improvement, so you learn to embrace it.

**Can you suggest a valuable "try this" for students considering a career in your profession?**
We all have that product or application that we hate to use. Find something you think could be better. Ask other people how they feel about it. Identify a problem for you to solve. Write down as many solutions you can think of. Sketch out your favorite. Test it. Refine it. There's no better way to learn than to just put it in practice. You can go to school, or go online to learn about it, but unless you're putting it to practice, you're not going to hone those skills. Nothing beats observing a problem, trying to understand it, and trying to solve it with other people.

*This interview was originally published in 2021.*

## *Conversation With...*

## JAY CORREIA

Founder and CEO
DreamCo Design, West Dundee, Illinois
Company owner, 11 years

**What was your individual career path in terms of education/training, entry-level job, or other significant opportunity?**

Simply put, I learned from experience. Every single day provides ample opportunity to learn. I didn't go to college. When I was 18 I wasn't 100 percent certain what, exactly, I wanted to do. I was leaning toward film school but a cosigner on a loan fell through and it never went any further. I contemplated other options but decided it made more sense financially to save money and learn via experience than go into debt without a specific career focus.

My first career-oriented position was with a digital printing company. I worked part-time during high school and though I began doing nothing more than making color copies, I ended up learning skills I still use to this day like Adobe Photoshop for graphic work.

I started DreamCo Design in 2006. After spending multiple years in the print/graphic design field, including working for a start-up, I came to the pragmatic realization that things were going digital. I was a self-taught/intern-taught graphic designer, a computer person, and full of ambition. I'd designed/developed a couple of successful websites and also excelled in sales related roles so I decided to fill a void by providing custom websites.

Within a couple of years, we reached the 7-figure mark for revenue. Our nationally recognized web/app/marketing agency has completed more than 1,500 projects for clients including Grammy Award winning musicians, TV personalities, the National Park Service, and a wide variety of small businesses.

In addition to marketing and web development, our company is, for example, currently developing social media platforms and apps. We've also done smaller networks that operate in a way similar to Facebook.

**What are the most important skills and/or qualities for someone in your profession?**

When being involved with social media, I think it's incredibly important to think like a business owner or decision maker. Social media is used as a reflection of an individual's or business' likeness. Having strong communication skills to extrapolate goals and deliver them digitally is a necessity.

**What do you wish you had known going into this profession?**
That no two clients are alike. While that may seem obvious, the amount of variance between the control some owners want to have as compared to the passiveness of others is striking.

**Are there many job opportunities in your profession? In what specific areas?**
There is a plethora of jobs in the social media industry.

Just as having a professional website is critical for most businesses, so is having a social media presence, especially for business-to-consumer (B2C) companies.

**How do you see your profession changing in the next five years? How do you see platforms evolving? What skill sets will be required?**
I think the concept of social media management will continue to grow, reaching maturity over the next decade.

Entrepreneurs are always dreaming up ideas that often involve social media. Privacy, issues involving who owns content, and users being paid for participating (whether real money, cryptocurrency, or other methods) will all likely play major roles in how the platforms evolve.

As basic as it sounds, having a general business understanding already helps. Social media management is a hybrid of putting technology to use for sales, branding, marketing, and public relations. In some cases, it is also customer service. Having a strong technological backbone will always be key.

**What do you enjoy most about your job? What do you enjoy least about your job?**
On the social media side, I enjoy seeing a campaign directly impact the success of a business or individual in a positive way. I least enjoy having to deal with unrealistic expectations. Not everything will go viral, no matter how much someone may want it to.

**Can you suggest a valuable "try this" for a student considering a career in your profession?**
Pick a company that you care about and develop a social media plan based around a hypothetical goal. For example, the goal might be "L.L.Bean wants to sell more boots while proving their boots outlast all of their competitors." With the goal in mind, choose the social media platforms you'd use to connect with consumers and plan the content accordingly. Think about every phase of carrying out the social media campaign. How are you going to get the content (photos, videos, etc.)? Are you going to schedule posts in advance? When are you going to post material? What demographics are you going to try to reach? How are you going to ensure that your strategy will help the company reach its goals?

Thinking through all these questions and putting a plan to paper is a great exercise that can also be hands on.

In addition to that, getting an internship with any marketing agency would also assist.

*This interview was originally published in 2018.*

## *Conversation With...*
## CRAIG KAHN

Software Developer, 34 years

**What was your individual career path in terms of education, entry-level job, or other significant opportunity?**

Although I have a bachelor's of arts degree, my education did not include any courses in computer science. I wish it had and I certainly recommend taking as many courses in computer science as possible if one's desired profession is to write software code. My education in software programming was strictly "on the job" at first, although I did take classes eventually. My first foray into programming came out of necessity. I was at a company that ran ski trips, via bus, to ski resorts in New England. We were dealing with multiple groups on multiple buses going to multiple resorts utilizing multiple rooms and types of lodging. In addition, some individuals needed rentals and/or lessons. This was 1979 and we were keeping track of all of this information by hand. The PC hadn't been invented yet (MS-DOS, the first popular operating system, had yet to be developed) but the desktop computer existed. It was actually the size of a desktop. Our company bought a dual 8" floppy disk drive computer running the CPM operating system. I learned how to program using a database program called dBase II.

**Are there many job opportunities in your profession? In what specific areas?**

I get the sense that there is a significant demand for programmers who can do more than merely write code. People are looking for programmers who can develop, or help develop, a complete application, including designing the user interface and developing the logic necessary for the program to work. The pure "coding" work seems to be going overseas.

**What do you wish you had known going into this profession?**

I wish I had taken classes and was more informed—there would have been less trial and error and I would have been up to speed much more quickly.

**How do you see your profession changing in the next five years?**

This is a tough one to call. My niche is developing small- to medium-scale applications for clients that want to move from keeping track of things on paper to being able to enter, retrieve and report on that information via computer. I think this will pretty much stay the same.

**What role will technology play in those changes, and what skills will be required?**

It's difficult to predict how the computer programming profession will change in the next five years. A lot will depend on the economy and the political environment, especially in regards to outsourcing. The good news is that certain types of programming cannot be shipped overseas because they require a lot of face-to-face meetings to go over and revise specifications. Basically, any development that would be considered "Agile Software Development," where requirements and solutions evolve over time as a result of working with others, would most likely need to be local.

As technology evolves, different skill sets will be needed. For example, 10 years ago a programmer didn't need to know much about web programming. These days it is essential to know how to develop for the desktop and for the web. I suspect that as smart phones become more powerful, and as tablets become more common, a requirement may be for an application to work among all platforms.

**Do you have any general advice or additional professional insights to share with someone interested in your profession?**

The most important requirement for a software developer is to be aware of what is going on in the world of technology and embrace the advances and changes. Even if you don't deem it necessary to learn how to use the newest technology, you should at least know it's out there and be ready to absorb and learn at a moment's notice.

Also, one of the things clients look for is someone with good communication skills—including being a good listener. If the choice is between two equally good coders, the one who the client feels can "get it"—and can explain what he or she is doing—will get the job.

**Can you suggest a valuable "try this" for students considering a career in your profession?**

I would suggest coming up with an idea for a program and spending time developing it in one's language of choice. Make mistakes and learn on your own time instead of on the client's time. If a potential client wants to know what you've developed in the past, you'll have something to show them.

*This interview was originally published in 2015.*

# Tax Examiner/ Collector/Revenue Agent

## Snapshot

**Career Cluster(s):** Finance

**Interests:** Accounting; Finance; Data; Statistics; Recordkeeping; Investigating

**Earnings (Yearly Average):** $56,780

**Employment & Outlook:** Decline In Growth Expected

## OVERVIEW

### Sphere of Work

Tax examiners and collectors, and revenue agents determine how much is owed in taxes and collect tax from individuals and businesses on behalf of federal, state, and local governments. They review tax returns, conduct audits, identify taxes owed, and collect overdue tax payments. Because of the sensitive financial information they deal in, it is a top concern for these workers to protect their systems from crimes such as ransomware, spyware, computer viruses, phishing, vishing, spoofing, data breaches, and more.

Photo via iStock.com/Andrew_Rybalko. [Used under license.]

## Work Environment

Tax examiners and collectors, and revenue agents work primarily in an office setting; others spend most of their time conducting field audits in taxpayers' homes or places of business.

### Profile

**Working Conditions:** Inside
**Education Needs:** Bachelor's Degree
**Licensure/Certification:** Not Required
**Opportunities for Experience:** Internship
**Interest Score:** CE

## Occupation Interest

Tax examiners and collectors, and revenue agents are both conventional and enterprising in nature, meaning they excel in this profession where following a set of procedures and routines is the standard, but they also make good leaders and are capable of making many decisions.

## A Day in the Life—Duties and Responsibilities

Tax examiners and collectors, and revenue agents ensure that individuals and businesses pay the appropriate amount of taxes owed, as prescribed by laws and regulations. In addition to verifying that tax returns are filed correctly, they follow up with taxpayers whose returns are questionable or who owe more money.

### Duties and Responsibilities

- Reviewing filed tax returns
- Determining whether credits and deductions claimed are allowed by law
- Contacting taxpayers to address problems and to request supporting documentation
- Conducting field audits and investigations of income tax returns
- Verifying information or updating tax liabilities
- Evaluating financial information, using their understanding of accounting procedures and knowledge of changes to tax laws and regulations
- Keeping records on each active case including telephone numbers and actions taken
- Notifying taxpayers of overpayment or underpayment and issuing a refund or requesting additional payment

Different levels of government collect different types of taxes. The federal government's Internal Revenue Service (IRS) deals primarily with business and personal income taxes. State governments collect income and sales taxes. Local governments collect property and sales taxes. At all levels, following the IRS cybersecurity recommendations for tax professionals is a necessity to keep information secure.

Because some states base income taxes on taxpayers' reported federal income, tax examiners working for the federal government report to the states any adjustments or corrections they make. State tax examiners then determine whether the adjustments affect the state taxpayer liability.

## OCCUPATION SPECIALTIES

**Revenue Agent**

Revenue agents, called internal revenue agents in the IRS, specialize in tax-related accounting. Like tax examiners, they review returns for accuracy. However, revenue agents handle the complex tax returns of large businesses and corporations.

Some experienced revenue agents focus exclusively on a particular area, such as multinational business. Regardless of their specialty, revenue agents must keep up to date with changes in tax laws and regulations.

**Tax Collector**

Tax collectors, also called internal revenue officers in the IRS, deal with overdue accounts. The process of collecting an overdue payment starts with the revenue agent or tax examiner sending a report to the taxpayer. If the taxpayer makes no effort to pay, the case is assigned to a tax collector.

When a tax collector takes a case, he or she first sends a notice to the taxpayer. The tax collector then works with the taxpayer to settle the debt. Settlement may involve setting up a plan in which the amount owed is paid back in small amounts over time.

When delinquent taxpayers claim that they cannot pay their taxes, collectors investigate and verify the claims. Tax collectors research information on taxpayer financial statements or mortgages and locate taxpayer-owned items of value through third parties, such as local departments of motor vehicles. Ultimately, they must decide whether the IRS should place a lien—a claim on an asset such as a bank account, real estate, or an automobile—to settle a debt. Tax collectors also have the authority to garnish wages—that is, take a portion of earned wages—to collect taxes owed.

**Tax Examiner**

Tax examiners typically deal with simple tax returns filed by individual taxpayers who claim few deductions and by small businesses. Tax examiners also may contact individual taxpayers in order to resolve outstanding problems with their returns.

Much of a tax examiner's job involves making sure that tax credits and deductions claimed by taxpayers are lawful. If a taxpayer owes additional taxes, tax examiners adjust the total amount by assessing fees, interest, and penalties and then notify the taxpayer of the total amount owed.

## WORK ENVIRONMENT

### Immediate Physical Environment

Most tax examiners and collectors work in offices, but some work from home. Although examiners and collectors usually work in teams, some work alone.

### Human Environment

Examiners and collectors interact with the public regularly; as such, they must be comfortable dealing with people. Many times, they will have to explain complex details to the public in a clear fashion, so their communication skills need to be strong. When pursuing overdue accounts, tax collectors should be firm and composed.

### Technological Environment

Examiners and collectors work with various software programs in the course of their work, such as accounting and bookkeeping software.

## EDUCATION AND TRAINING

### High School/Secondary

High school students interested in becoming tax examiners, collectors or revenue agents should focus on math, statistics, and business-related courses. This job will require at least a bachelor's degree, so students should prepare themselves for postsecondary education, including researching the schools they would like to consider attending to achieve that goal.

**Suggested High School Subjects**

- Algebra
- Civics
- Economics
- English
- Pre-Calculus
- Statistics

**Related Career Pathways/Majors**

*Finance Career Cluster*

- Banking Services Pathway
- Business Finance Pathway
- Securities & Investments Pathway

### Postsecondary

Tax examiners and collectors, and revenue agents typically need a bachelor's degree in accounting or a related field. However, the required level of education and experience varies with the position and employer.

Tax examiners and collectors, and revenue agents typically need a bachelor's degree in accounting or a related field, such as business. For some jobs, work experience may substitute for a degree.

Candidates for tax examiner and collector positions at the Internal Revenue Service (IRS) may qualify with a bachelor's degree in any field of study or with specialized experience, or with a combination of education and experience. Internal revenue agents at the IRS generally need a bachelor's degree in accounting; a combination of education and experience equivalent to a major in accounting; or a Certified Public Accountant (CPA) certificate.

### *Transferable Skills and Abilities*

**Analytical Skills**

- Identifying questionable claims for credits and deductions
- Determining if claims are lawful

**Communication Skills**

- Clearly explaining complex details, especially about sensitive information, in their work with the public

**Detail-oriented**

- Paying attention to details
- Verifying the accuracy of each entry on the tax returns they review

**Interpersonal Skills**

- Being at ease interacting with people
- Remaining firm and composed when pursuing overdue accounts

**Math Skills**

- Dealing with numbers daily and being comfortable with arithmetic
- Analyzing, comparing, and interpreting facts and figures

**Organizational Skills**

- Working with multiple returns and a variety of financial documents

Newly hired tax examiners and collectors, and revenue agents typically receive on-the-job training that lasts between 1 month and 1 year. These workers also must keep current with changes to the tax code and in enforcement and collection procedures.

### Related College Majors

- Accounting
- Auditing
- Business Management & Administration
- Economics
- Financial Management
- General Business

### Adult Job Seekers

Some employers accept work experience as a substitute for education. For example, employers may hire tax examiners and revenue agents who have experience as accountants or bookkeepers, or they may hire tax collectors who have experience working as bill and account collectors, customer service representatives, or credit checkers.

## EARNINGS AND ADVANCEMENT

The median annual wage for tax examiners and collectors, and revenue agents was $56,780 in 2021. The lowest 10 percent earned less than $35,260, and the highest 10 percent earned more than $102,840.

Tax examiners who review individual tax returns may advance to revenue agent positions, working on more complex business returns.

Tax examiners and collectors, and revenue agents who demonstrate leadership skills and a thorough knowledge of tax collection activities may advance to supervisory or managerial positions.

## EMPLOYMENT AND OUTLOOK

Tax examiners and collectors, and revenue agents held about 55,300 jobs in 2021. Employment is projected to decline 7 percent from 2021 to 2031.

Despite declining employment, about 4,300 openings for tax examiners and collectors, and revenue agents are projected each year, on average, over the decade. All of those openings are expected to result from the need to replace workers who transfer to other occupations or exit the labor force, such as to retire.

Employment of these workers will depend primarily on future changes to federal, state, and local government budgets. Because the salaries of tax examiners and collectors, and revenue agents are paid with public funds, budgetary constraints may directly impact governments' ability to hire these workers and, in turn, affect employment growth.

**Fast Fact**

**In 2016, it was estimated that between $1.68 billion to $2.31 billion of federal income tax refunds reached identity thieves (including cybercriminals), despite the defensive measures employed by the IRS. An estimated 740,000 to 810,000 tax returns were compromised.** Source: Bloombergtax.com

Photo via iStock.com/Douglas Rissing. [Used under license.]

### Related Occupations

- Accountant/Auditor
- Bookkeeping/Accounting/Auditing Clerks
- Budget Analyst
- Cost Estimator
- Financial Analyst
- Financial Examiner
- Financial Manager
- Human Resources Specialist
- Loan Officer
- Personal Financial Advisor

## MORE INFORMATION

**IRS—Careers**
www.jobs.irs.gov

**U.S. Office of Personnel Management**
www.opm.gov/policy-data-oversight/classification-qualifications/general-schedule-qualification-standards/#url=0500-ndx

*Daniel Vest*

# What Are Your Career Interests?

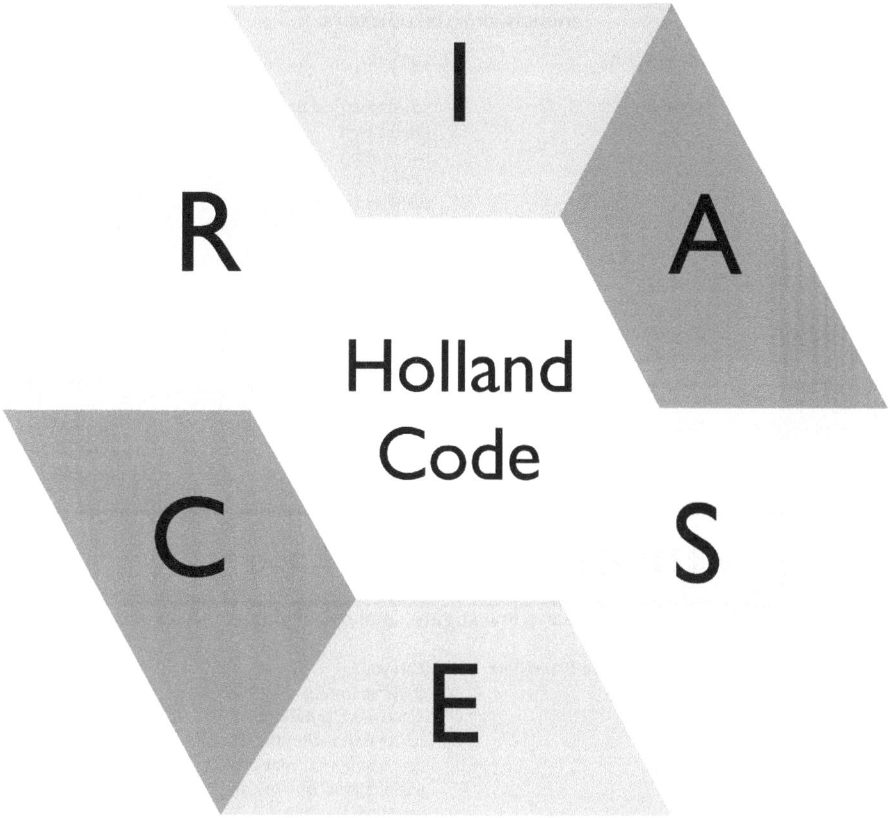


This is based on Dr. John Holland's theory that people and work environments can be loosely classified into six different groups. Each of the letters above corresponds to one of the six groups described in the following pages.

Different people's personalities may find different environments more to their liking. While you may have some interests in and similarities to several of the six groups, you may be attracted primarily to two or three of the areas. These two or three letters are your "Holland Code." For example, with a code of "RES" you would most resemble the Realistic type, somewhat less resemble the Enterprising type, and resemble the Social type even less. The types that are not in your code are the types you resemble least of all.

Most people, and most jobs, are best represented by some combination of two or three of the Holland interest areas. In addition, most people are most satisfied if there is some degree of fit between their personality and their work environment.

The rest of the pages in this booklet further explain each type and provide some examples of career possibilities, areas of study at MU, and co-curricular activities for each code. To take a more in-depth look at your Holland Code, take a self-assessment such as the SDS, Discover, or a card sort at the MU Career Center with a Career Specialist.

# Realistic (Doers)

**People who have athletic ability, prefer to work with objects, machines, tools, plants or animals, or to be outdoors.**

**Are you?**
practical
straightforward/frank
mechanically inclined
stable
concrete
reserved
self-controlled
independent
ambitious
systematic

Can you?
fix electrical things
solve electrical problems
pitch a tent
play a sport
read a blueprint
plant a garden
operate tools and machine

Like to?
tinker with machines/vehicles
work outdoors
be physically active
use your hands
build things
tend/train animals
work on electronic equipment

**Career Possibilities (Holland Code):**

Air Traffic Controller (SER)
Archaeologist (IRE)
Athletic Trainer (SRE)
Cartographer (IRE)
Commercial Airline Pilot (RIE)
Commercial Drafter (IRE)
Corrections Officer (SER)
Dental Technician (REI)
Farm Manager (ESR)
Fish and Game Warden (RES)
Floral Designer (RAE)
Forester (RIS)
Geodetic Surveyor (IRE)
Industrial Arts Teacher (IER)
Laboratory Technician (RIE)
Landscape Architect (AIR)
Mechanical Engineer (RIS)
Optician (REI)
Petroleum Geologist (RIE)
Police Officer (SER)
Practical Nurse (SER)
Property Manager (ESR)
Recreation Manager (SER)
Service Manager (ERS)
Software Technician (RCI)
Ultrasound Technologist (RSI)
Vocational Rehabilitation Consultant (ESR)

# Investigative (Thinkers)

**People who like to observe, learn, investigate, analyze, evaluate, or solve problems.**

Are you?
inquisitive
analytical
scientific
observant/precise
scholarly
cautious
intellectually self-confident
Independent
logical
complex
Curious

**Can you?**
think abstractly
solve math problems
understand scientific theories
do complex calculations
use a microscope or computer
interpret formulas

Like to?
explore a variety of ideas
work independently
perform lab experiments
deal with abstractions
do research
be challenged

**Career Possibilities (Holland Code):**

Actuary (ISE)
Agronomist (IRS)
Anesthesiologist (IRS)
Anthropologist (IRE)
Archaeologist (IRE)
Biochemist (IRS)
Biologist (ISR)
Chemical Engineer (IRE)
Chemist (IRE)
Computer Systems Analyst (IER)
Dentist (ISR)
Ecologist (IRE)
Economist (IAS)
Electrical Engineer (IRE)
Geologist (IRE)
Horticulturist (IRS)
Mathematician (IER)
Medical Technologist (ISA)
Meteorologist (IRS)
Nurse Practitioner (ISA)
Pharmacist (IES)
Physician, General Practice (ISE)
Psychologist (IES)
Research Analyst (IRC)
Statistician (IRE)
Surgeon (IRA)
Technical Writer (IRS)
Veterinarian (IRS)

# Artistic *(Creators)*

**People who have artistic, innovating, or intuitional abilities and like to work in unstructured situations using their imagination and creativity.**

**Are you?**
creative
imaginative
innovative
unconventional
emotional
independent
Expressive
original
introspective
impulsive
sensitive
courageous
complicated
idealistic
nonconforming

**Can you?**
sketch, draw, paint
play a musical instrument
write stories, poetry, music
sing, act, dance
design fashions or interiors

**Like to?**
attend concerts, theatre, art exhibits
read fiction, plays, and poetry
work on crafts
take photography
express yourself creatively
deal with ambiguous ideas

**Career Possibilities (Holland Code):**

Actor (AES)
Advertising Art Director (AES)
Advertising Manager (ASE)
Architect (AIR)
Art Teacher (ASE)
Artist (ASI)
Copy Writer (ASI)
Dance Instructor (AER)
Drama Coach (ASE)
English Teacher (ASE)
Entertainer/Performer (AES)
Fashion Illustrator (ASR)
Interior Designer (AES)
Intelligence Research Specialist (AEI)
Journalist/Reporter (ASE)
Landscape Architect (AIR)
Librarian (SAI)
Medical Illustrator (AIE)
Museum Curator (AES)
Music Teacher (ASI)
Photographer (AES)
Writer (ASI)
Graphic Designer (AES)

# Social *(Helpers)*

**People who like to work with people to enlighten, inform, help, train, or cure them, or are skilled with words.**

**Are you?**
friendly
helpful
idealistic
insightful
outgoing
understanding
cooperative
generous
responsible
forgiving
patient
kind

**Can you?**
teach/train others
express yourself clearly
lead a group discussion
mediate disputes
plan and supervise an activity
cooperate well with others

**Like to?**
work in groups
help people with problems
do volunteer work
work with young people
serve others

**Career Possibilities (Holland Code):**

City Manager (SEC)
Clinical Dietitian (SIE)
College/University Faculty (SEI)
Community Org. Director (SEA)
Consumer Affairs Director (SER)Counselor/Therapist (SAE)
Historian (SEI)
Hospital Administrator (SER)
Psychologist (SEI)
Insurance Claims Examiner (SIE)
Librarian (SAI)
Medical Assistant (SCR)
Minister/Priest/Rabbi (SAI)
Paralegal (SCE)
Park Naturalist (SEI)
Physical Therapist (SIE)
Police Officer (SER)
Probation and Parole Officer (SEC)
Real Estate Appraiser (SCE)
Recreation Director (SER)
Registered Nurse (SIA)
Teacher (SAE)
Social Worker (SEA)
Speech Pathologist (SAI)
Vocational-Rehab. Counselor (SEC)
Volunteer Services Director (SEC)

# Enterprising *(Persuaders)*

**People who like to work with people, influencing, persuading, leading or managing for organizational goals or economic gain.**

**Are you?**
self-confident
assertive
persuasive
energetic
adventurous
popular
ambitious
agreeable
talkative
extroverted
spontaneous
optimistic

**Can you?**
initiate projects
convince people to do things your way
sell things
give talks or speeches
organize activities
lead a group
persuade others

**Like to?**
make decisions
be elected to office
start your own business
campaign politically
meet important people
have power or status

**Career Possibilities (Holland Code):**

Advertising Executive (ESA)
Advertising Sales Rep (ESR)
Banker/Financial Planner (ESR)
Branch Manager (ESA)
Business Manager (ESC)
Buyer (ESA)
Chamber of Commerce Exec (ESA)
Credit Analyst (EAS)
Customer Service Manager (ESA)
Education & Training Manager (EIS)
Emergency Medical Technician (ESI)
Entrepreneur (ESA)
Foreign Service Officer (ESA)
Funeral Director (ESR)
Insurance Manager (ESC)
Interpreter (ESA)
Lawyer/Attorney (ESA)
Lobbyist (ESA)
Office Manager (ESR)
Personnel Recruiter (ESR)
Politician (ESA)
Public Relations Rep (EAS)
Retail Store Manager (ESR)
Sales Manager (ESA)
Sales Representative (ERS)
Social Service Director (ESA)
Stockbroker (ESI)
Tax Accountant (ECS)

# Conventional *(Organizers)*

**People who like to work with data, have clerical or numerical ability, carry out tasks in detail, or follow through on others' instructions.**

**Are you?**
well-organized
accurate
numerically inclined
methodical
conscientious
efficient
conforming
practical
thrifty
systematic
structured
polite
ambitious
obedient
persistent

**Can you?**
work well within a system
do a lot of paper work in a short time
keep accurate records
use a computer terminal
write effective business letters

**Like to?**
follow clearly defined procedures
use data processing equipment
work with numbers
type or take shorthand
be responsible for details
collect or organize things

**Career Possibilities (Holland Code):**

Abstractor (CSI)
Accountant (CSE)
Administrative Assistant (ESC)
Budget Analyst (CER)
Business Manager (ESC)
Business Programmer (CRI)
Business Teacher (CSE)
Catalog Librarian (CSE)
Claims Adjuster (SEC)
Computer Operator (CSR)
Congressional-District Aide (CES)
Cost Accountant (CES)
Court Reporter (CSE)
Credit Manager (ESC)
Customs Inspector (CEI)
Editorial Assistant (CSI)
Elementary School Teacher (SEC)
Financial Analyst (CSI)
Insurance Manager (ESC)
Insurance Underwriter (CSE)
Internal Auditor (ICR)
Kindergarten Teacher (ESC)
Medical Records Technician (CSE)
Museum Registrar (CSE)
Paralegal (SCE)
Safety Inspector (RCS)
Tax Accountant (ECS)
Tax Consultant (CES)
Travel Agent (ECS)

# GENERAL BIBLIOGRAPHY


Anderson, Nate. *The Internet Police: How Crime Went Online and the Cops Followed.* New York: W.W. Norton, 2014.

Antonucci, Domenic. *The Cyber Risk Handbook: Creating and Measuring Effective Cybersecurity Capabilities.* Hoboken: Wiley, 2017.

Auger, Gerald, et. al. *Cybersecurity Career Master Plan: Proven Techniques and Effective Tips to Help You Advance in Your Cybersecurity Career.* Birmingham, UK: Packt Publishing, 2021.

Ayala, Luis. *Cybersecurity for Hospitals and Healthcare Facilities: A Guide to Detection and Prevention.* New York: Apress, 2016.

Barker, Jessica. *Confident Cyber Security: How to Get Started in Cyber Security and Future Proof Your Career.* London: Kogan Page, 2020.

Bennett, Daniel T., Paul L. Goethals, and Natalie M. Scala. *Mathematics in Cyber Research.* Boca Raton, FL: CRC Press, 2022.

Brunsman, Joseph E., Daniel W. Hudson, and Kenneth J. Reiners. *Damage Control: Cyber Insurance and Compliance.* Annapolis, MD: Chesapeake Professional Liability Brokers, 2020.

Carrascosa, Iván Palomares, Harsha Kumara Kalutarage, and Yan Huang, eds. *Data Analytics and Decision Support for Cybersecurity: Trends, Methodologies and Applications.* Cham, Switzerland: Springer, 2017.

Clinton, Larry, ed. *Cybersecurity for Business: Organization-Wide Strategies to Ensure Cyber Risk Is Not Just an IT Issue.* London: Kogan Page, 2022.

Clark, Colby. *Cybersecurity Incident Management Guide, Vol. 1: Preparation, Threat Response, & Post-Incident Activity.* Amazon KDP, 2019.

Codings, Zach. *Computer Programming and Cyber Security for Beginners.* CreateSpace, 2019.

Corradini, Isabella. *Building a Cybersecurity Culture in Organizations: How to Bridge the Gap Between People and Digital Technology.* Cham, Switzerland: Springer, 2021.

Diogenes, Yuri, and Dr. Erdal Ozkaya. *Cybersecurity - Attack and Defense Strategies.* 3rd ed. Birmingham, UK: Packt Publishing, 2022.

Espinosa, Christian. *The Smartest Person in the Room: The Root Cause and New Solution for Cybersecurity.* Austin, TX: Lioncrest Publishing, 2021.

Garber, Leeza, and Scott Olson. *Can. Trust. Will.: Hiring for the Human Element in the New Age of Cybersecurity.* New York: Business Expert Press, 2022.

Grubb, Sam. *How Cybersecurity Really Works: A Hands-On Guide for Total Beginners*. San Francisco: No Starch Press, 2021.

Harvard Business Review. *Cybersecurity: The Insights You Need.* Boston: Harvard Business Review Press, 2019.

Hayes, Darren. *A Practical Guide to Digital Forensics Investigations.* 2nd ed. Hoboken, NJ: Pearson, 2020.

Kosseff, Jeff. *Cybersecurity Law.* 3rd ed. Hoboken, NJ: Wiley, 2022.

Kovacs, Attila. *Cybersecurity for Beginners: How to Get a Job in Cybersecurity.* CreateSpace, 2019.

Leirvik, Ryan. *Understand, Manage, and Measure Cyber Risk: Practical Solutions for Creating a Sustainable Cyber Program.* New York: Apress, 2021.

Le-Khac, Nhien-An, and Kim-Kwang Raymond Choo, eds. *Cyber and Digital Forensic Investigations: A Law Enforcement Practitioner's Perspective.* Cham, Switzerland: Springer, 2020.

Limoncelli, Thomas, Christina J. Hogan, and Strata Chalup. *The Practice of System and Network Administration.* 3rd ed. Boston: Addison-Wesley, 2017.

Lopp, Michael. *The Software Developer's Career Handbook: A Guide to Navigating the Unpredictable.* Sebastopol, CA: O'Reilly, 2023.

Marion, Nancy E., and Jason Twede. *Cybercrime: An Encyclopedia of Digital Crime.* Santa Barbara: ABC-CLIO, 2020.

McHale, David. *The CPA Cybersecurity Survival Guide: Your Easy Guide to Spot Social Engineering and Phishing, Stop Ransomware and Fraud, and Sleep Soundly at Night.* Amazon KDP, 2020.

Miller, Alyssa. *Cybersecurity Career Guide.* Shelter Island, NY: Manning Publications, 2022.

Mongeau, Scott, and Andrzej Hajdasinski. *Cybersecurity Data Science: Best Practices in an Emerging Profession.* Cham, Switzerland: Springer, 2021.

Moseley, Ralph. *Advanced Cybersecurity Technologies.* Boca Raton, FL: CRC Press, 2021.

Oettinger, William. *Learn Computer Forensics: A Beginner's Guide to Searching, Analyzing, and Securing Digital Evidence.* Birmingham, UK: Packt Publishing, 2020.

Ozkaya, Erdal, and Milad Aslaner. *Hands-On Cybersecurity for Finance: Identify Vulnerabilities and Secure Your Financial Services from Security Breaches.* Birmingham, UK: Packt Publishing, 2019.

Patton, Helen E. *Navigating the Cybersecurity Career Path.* Hoboken: Wiley, 2022.

Spitters, Thomas H. *A Cybersecurity Breviary for Accountants.* Baume Verlag, 2017.

Underhill, Ken, et. al. *Hack the Cybersecurity Interview: A Complete Interview Preparation Guide for Jumpstarting Your Cybersecurity Career.* Birmingham, UK: Packt Publishing, 2022.

Wolff, Josephine. *Cyberinsurance Policy: Rethinking Risk in an Age of Ransomware, Computer Fraud, Data Breaches, and Cyberattacks.* Cambridge, MA: The MIT Press, 2022.

# LIST OF ORGANIZATIONS AND RESOURCES

**Academic Data Science Alliance**
www.academicdatascience.org
Alliance for Data Science Professionals
info@afdsp.co.uk
www.alliancefordatascienceprofessionals.com

**Advance CTE: State Leaders Connecting Learning to Work**
8484 Georgia Avenue, Suite 620
Silver Spring, MD 20910
301.588.9630
careertech.org

**American Academy of Forensic Sciences (AAFS)**
410 North 21st Street
Colorado Springs, CO 80904
719.636.1100
info@aafs.org
www.aafs.org

**American Academy of Professional Coders (AAPC)**
2233 S Presidents Drive Suite F
Salt Lake City, UT 84120
800.626.2633
www.aapc.com

**American Accounting Association (AAA)**
9009 Town Center Parkway
Lakewood Ranch, FL 34202
941.921.7747
info@aaahq.org
aaahq.org
Forensic Accounting Division: aaahq.org/FIA

**American Association for Budget and Program Analysis (AABPA)**
P.O. Box 1157
Falls Church, VA 22041
AABPA@aabpa.org
www.aabpa.org

**American Bankers Association (ABA)**
1120 Connecticut Avenue NW
Washington, DC 20036
800.226.5377
www.aba.com

**American Board of Criminalistics**
P.O. Box 1358
Palmetto, FL 34220
941.729.9050
acbcregistrar@criminalistics.com
www.criminalistics.com

**American Board of Forensic Accounting (ABFA)**
1400 Easton Drive, Suite 137
Bakersfield, CA 93309
661.861.8861
staff@abfa.us

**American Board of Medicolegal Death Investigators (ABMDI)**
10104 Senate Drive, Suite 241
Lanham, MD 20706
410.807.3007
info@abmdi.org
www.abmdi.org

**American Health Information Management Association (AHIMA)**
233 N. Michigan Avenue, 21st Floor
Chicago, IL 60601
800.335.5535
info@ahima.org
www.ahima.org

**American Institute of Certified Public Accountants (AICPA)**
1345 Avenue of the Americas, 27th Floor
New York, NY 10105
212.596.6200
www.aicpa.org

**American Institution of Certified Public Accountants (AICPA)**
220 Leigh Farm Road
Durham, NC 27707-8110
888.777.7077
www.aicpa.org

**American Mathematical Society (AMS)**
201 Charles Street
Providence, RI 02904-2213
800.321.4267
www.ams.org

**American Society for Quality (ASQ)**
P.O. Box 3005
Milwaukee, WI 53201-3005
800.248.1946
asq.org

**American Statistical Association (ASA)**
732 North Washington Street
Alexandria, VA 22314-1943
888.231.3473
asainfo@amstat.org
www.amstat.org

**ASIS International**
1625 Prince Street
Alexandria, VA 22314-2882
703.519.6200
asis@asisonline.org

**Association for Computing Machinery (ACM)**
1601 Broadway, 10th Floor
New York, NY 10019-7434
212.869.7440
acmhelp@acm.org
www.acm.org

**Association for Financial Professionals (AFP)**
4520 East-West Highway, Suite 800
Bethesda, MD 20814
301.907.2862
customerservice@AFPonline.org
www.afponline.org

**Association for Software Testing (AST)**
marketing@associationforsoftwaretesting.org
associationforsoftwaretesting.org

**Association for Women in Computing (AWC)**
P.O. Box 2768
Oakland, CA 94602
info@awc-hq.org
www.awc-hq.org

**Association of Certified Fraud Examiners (ACFA)**
Global Headquarters—The Gregor Building
716 West Avenue
Austin, TX 78701-2727
800.245.3321
MemberServices@ACFE.com
www.acfe.com

**Association of Firearm and Tool Mark Examiners (AFTE)**
550 West 16th Street
Indianapolis, IN United States 46202
webmaster@afte.org
www.afte.org

**Association of Government Accountants (AGA)**
2208 Mount Vernon Avenue
Alexandria, VA 22301-1314
703.684.6931
www.agacgfm.org

**Association of Information Technology Professionals (AITP)**
3500 Lacey Road, Suite 100
Downers Grove, IL 60515
866.835.8020
630.678.8300
www.comptia.org

**Association of International Certified Professional Accountants (AICPA)**
220 Leigh Farm Road
Durham, NC 27707-8110
888.777.7077
www.aicpa.org

**Association of Support Professionals (ASP)**
1401 West Grand Avenue
Hot Springs, AK 71913
503.780.3520
info@asponline.com
www.asponline.com

**Association to Advance Collegiate Schools of Business (AACSB)**
777 S. Harbour Island Boulevard, Suite 750
Tampa, FL 33602
813.769.6500
www.aacsb.edu

**Bureau of Alcohol, Tobacco, Firearms and Explosives (ATF)**
99 New York Avenue NE
Washington, DC 20226
202.648.7080
www.atf.gov

**Center for Statistics and Applications in Forensic Evidence (CSAFE)**
195 Durham Center
613 Morrill Road
Ames, IA 50011
www.forensicstats.org

**Certified Financial Planner Board of Standards (CFP)**
1425 K Street NW, Unit 800
Washington, DC 20005
800.487.1497
mail@cfpboard.org
www.cfp.net

**CFA Institute**
1401 New York Avenue NW, Suite 330
Washington, DC 20005
434.951.5499
info@cfainstitute.org
www.cfainstitute.org

**Commission on Accreditation for Health Informatics and Information Management Education (CAHIIM)**
200 East Randolph Street, Suite 5100
Chicago, IL 60601
312.235.3255
info@cahiim.org
www.cahiim.org

**CompTIA**
3500 Lacey Road, Suite 100
Downers Grove, IL 60515
866.835.8020
membership@comptia.org
www.comptia.org

**Computing Research Association (CRA)**
1828 L Street NW, Suite 800
Washington, DC 20036
202.234.2111
cra.org

**Computing Technology Industry Association (CompTIA)**
3500 Lacey Road, Suite 100
Downers Grove, IL 60515
866.835.8020
www.comptia.org

**Consumer Financial Protection Bureau (CFPB)**
1700 G Street NW
Washington, DC 20552
855.411.2372
www.consumerfinance.gov

**Council of International Investigators (CII)**
P.O. Box 1015
St. Helena, SC 29920
917.292.0613
lec@cii2.org
www.cii2.org

**Dictionary of Occupational Titles (DOT)**
occupationalinfo.org

**Exploring National Service Center**
1325 West Walnut Hill Lane
Irving, TX 75038
972.580.2433
exploring@lflmail.org
www.exploring.org
Offers the Law Enforcement Exploring program:
www.exploring.org/law-enforcement

**Federal Bureau of Investigation (FBI)**
935 Pennsylvania Avenue NW
Washington, DC 20535-0001
202.324.3000
www.fbi.gov

**Federal Deposit Insurance Corporation (FDIC)**
Washington, DC
877.275.3342
www.fdic.gov

**Financial Industry Regulatory Authority (FINRA)**
1735 K Street NW
Washington, DC 20006
301.590.6500
www.finra.org

**Forensic Accountants Society (FASNA)**
6200 Aurora Avenue, Suite 402W
Urbandale, IA 50322-2800
515.276.7000
www.fasna.org

**Forensic Certified Public Accountants Association (FCPAA)**
1120 Avenue of the Americas, Room 4057, 4th Floor
New York, NY 10036
626.521.9077
info@fcpachina.org
www.fcpai.org

**Forensic CPA Society**
P.O. Box 31060
Spokane, WA 99223
509.448.9318
www.fcpas.org

**Global Academy of Finance and Management (GAFM)**
P.O. Box 293
1670F E. Cheyenne Mountain Boulevard
Colorado Springs, CO 80906
www.gafm.com

**Government Finance Officers Association (GFOA)**
203 N. LaSalle Street, Suite 2700
Chicago, IL 60601
www.gfoa.org

**Help Desk Institute (HDI)**
121 South Tejon Street, Suite 1100
Colorado Springs, CO 80903
800.248.5667
www.thinkhdi.com

**HR Certification Institute (HRCI)**
1725 Duke Street, Suite 400
Alexandria, VA 22314
866.898.4724
info@hrci.org
www.hrci.org

**IEEE Computer Society**
2001 L Street NW, 700
Washington, DC 20036-4928
202.371.0101
help@computer.org
www.computer.org

**Information Systems Audit and Control Association (ISACA)**
1700 E. Golf Road, Suite 400
Schaumburg, IL 60173
847.253.1545
www.isaca.org

**Institute for Operations Research and the Management Sciences (INFORMS)**
5521 Research Park Drive, Suite 200
Catonsville, MD 21228
800.446.3676
informs@informs.org
www.informs.org

**Institute for the Certification of Computing Professionals (ICCP)**
244 South Randall Road, Suite 116
Elgin, IL 60123
800.843.8227
office2@iccp.org
www.iccp.org

**Institute of Certified Forensic Accountants (ICFA)**
3422 Old Capitol Trail, Suite 1063
Wilmington, DE 19808-6192
302.425.5528
www.forensicglobal.org

**Institute of Electrical and Electronics Engineers Computer Society (IEEE-CS)**
2000 L Street NW, Suite 700
Washington, DC 20036-4928
202.371.0101
help@computer.org
www.computer.org

**Institute of Internal Auditors (IIA)**
1035 Greenwood Boulevard, Suite 401
Lake Mary, FL 32746
407.937.1111
na.theiia.org

**Institute of Management Accountants (IMA)**
10 Paragon Drive, Suite 1
Montvale, NJ 07645-1760
800.638.4427
ima@imanet.org
www.imanet.org

**Institute of Management Consultants (IMC)**
2598 E. Sunrise Boulevard, Suite 2104
Fort Lauderdale, FL 33304
800.793.4992
www.imcusa.org

**Insurance Information Institute (III)**
110 William Street
New York, NY 10038
212.346.5500
info@iii.org
www.iii.org

**International Association of Computer Science and Information Technology (IACSIT)**
contact@iacsit.org
www.iacsit.org

**International Claim Association (ICA)**
1800 M Street, NW, 400 South
Washington, DC 20036
202.452.0143
memberservices@claim.org
www.claim.org

**International Crime Scene Investigators Association (ICSIA)**
P.O. Box 1532
Grand Rapids, MI 49501
800.504.7501
icsia@icsia.org
www.icsia.org

**International Foundation of Employee Benefit Plans (IFEBP)**
18700 W. Bluemound Road
Brookfield, WI 53045
888.334.3327
www.ifebp.org

**International Intelligence Network (INTELLENET)**
P.O. Box 350
Gladwyne, PA 19035
734.320.9240
Intellenet@intellenet.org
www.intellenet.org

**International Public Management Association for Human Resources (IPMA-HR)**
1617 Duke Street
Alexandria, VA 22314
703.549.7100
www.ipma-hr.org

**IRS-Careers**
www.jobs.irs.gov

**Military Operations Research Society (MORS)**
2111 Wilson Boulevard, Suite 700
Arlington, VA 22201
703.933.9070
www.mors.org

**Mortgage Bankers Association (MBA)**
1919 M Street NW, 5th floor
Washington, DC 20036
202.557.2700
membership@mba.org
www.mba.org

**National Association of Forensic Accountants (NAFA)**
6635 West Commercial Boulevard, Suite 217
Tamarac, FL 33319
844.217.6975
mail@nafanet.com
www.nafanet.com

**National Association of Insurance and Financial Advisors (NAIFA)**
2901 Telestar Court
Falls Church, VA 22042
877.866.2432
info@naifa.org
belong.naifa.org

**National Association of Legal Investigators (NALI)**
P.O. Box 278
Bath, MI 48808
517.702.9835
www.nalionline.org

**National Association of Mortgage Brokers (NAMB)**
601 Pennsylvania Avenue NW, South Building
Washington, DC 20004
202.434.8250
membership@namb.org
namb.org

**National Association of Mortgage Processors (NAMP)**
www.mortgageprocessor.org

**National Association of Mortgage Underwriters (NAMU)**
Presidential Circle
4000 Hollywood Boulevard, Suite 555-South
Hollywood, FL 33021
800.865.8550
contact@mortgage-underwriters.org
www.Mortgage-Underwriters.org

**National Association of Police Organizations (NAPO)**
317 South Patrick Street
Alexandria, VA 22314
703.549.0775
www.napo.org

**National Association of Professional Process Servers (NAPPS)**
P.O. Box 4547
Portland, OR 97208-4547
503.222.4180
napps.org

**National Association of Public Insurance Adjusters (NAPIA)**
1400 Village Square Blvd, Suite 3-187
Tallahassee, FL 32312
703.433.9217
info@napia.com
napia.com

**National Association of State Budget Officers (NASBO)**
Hall of States Building
444 North Capitol Street, NW Suite 642
Washington, DC 20001
202.624.5382
nasbo-direct@nasbo.org
www.nasbo.org

**National Center for Women & Information Technology (NCWIT)**
1909 26th Street, 2nd Floor
Boulder, CO 80302
303.735.6671
info@ncwit.org
ncwit.org

**National Council of Investigation and Security Services (NCISS)**
P.O. Box 200615
Evans, CO 80620
800.445.8408
www.nciss.org

**National Defender Investigator Association (NDIA)**
P.O. Box 1900
San Marcos, CA 92079-1900
760.290.3909
nationaldefender@gmail.com
www.ndia.net

**National Healthcareer Association (NHA)**
11161 Overbrook Road
Leawood, KS 66211
800.499.9092
www.nhanow.com

**National Insurance Producer Registry (NIPR)**
1100 Walnut Street, Suite 1500
Kansas City, MO 64106
www.nipr.com

**National Police Association (NPA)**
8710 Bash Street, Suite 501692
Indianapolis, IN 46250
302.469.1765
nationalpolice.org

**National Police Foundation (NPI)**
2550 S Clark Street, Suite 1130
Arlington, VA 22202
202.833.1460
info@policefoundation.org
www.policefoundation.org

**Nationwide Multistate Licensing System & Registry (NMLS) Resource Center**
855.665.7123
mortgage.nationwidelicensingsystem.org

**North American Securities Administrators Association (NASAA)**
750 First Street NE, Suite 990
Washington, DC 20002
202.737.0999
www.nasaa.org

**O*NET OnLine**
onet@onetcenter.org
www.onetonline.org
My Next Move: www.mynextmove.org

**Professional Background Screening Association (PBSA)**
110 Horizon Drive, Suite 210
Raleigh, NC 27615
919.459.2082
info@thepbsa.org
thepbsa.org

**Quality Assurance Association (QAA)**
1301 Riverplace Boulevard, Suite 800
Jacksonville, FL 32207
QAA@qualityassuranceassociation.com
www.qualityassuranceassociation.com

**Society for Human Resource Management (SHRM)**
1800 Duke Street
Alexandria, VA 22314
800.283.7476
shrm@shrm.org
www.shrm.org

**Society for Industrial and Applied Mathematics (SIAM)**
3600 Market Street, 6th Floor
Philadelphia, PA 19104
800.447.7426
www.siam.org

**Society of Financial Examiners (SOFE)**
3505 Vernon Woods Drive
Summerfield, NC 27358
800.787.7633
info@sofe.org
www.sofe.org

**Society of Quality Assurance (SQA)**
154 Hansen Road, Suite 201
Charlottesville, VA 22911
434.297.4772
sqa@sqa.org
www.sqa.org

**Technology & Services Industry Association (TSIA)**
17065 Camino San Bernardo, Suite 200
San Diego, CA 92127
858.674.5491
www.tsia.com

**The American College of Financial Services**
630 Allendale Road, Suite 400
King of Prussia, PA 19406
610.526.1000
866.216.8112
www.theamericancollege.edu

**The Institutes**
720 Providence Road, Suite 100
Malvern, PA 19355
800.644.2101
customersuccess@theinstitutes.org
www.theinstitutes.org

**USAJOBS**
c/o U.S. Office of Personnel Management
1900 E Street NW
Washington, DC 20415-1000
202.606.1800
www.usajobs.gov

**U.S. Bureau of Labor Statistics (BLS)**
Postal Square Building
2 Massachusetts Avenue NE
Washington, DC 20212-0001
202.691.5200
www.bls.gov
Occupational Outlook Handbook: www.bls.gov/ooh

**U.S. Customs and Border Protection (CBP)**
1300 Pennsylvania Avenue NW
Washington, DC 20229
202.325.8000
www.cbp.gov

**U.S. Department of Homeland Security (DHS)**
2707 Martin Luther King Jr. Avenue SE
Washington, DC 20528-0525
202.282.8000
www.dhs.gov

**U.S. Drug Enforcement Administration (DEA)**
600-700 Army Navy Drive
Arlington, VA 22202
202.307.1000
www.dea.gov

**U.S. Marshals Service (USMS)**
Arlington, VA
202.307.9100
www.usmarshals.gov

**U.S. Office of Personnel Management (OPM)**
1900 E Street, NW
Washington, DC 20415-1000
202.606.1800
www.opm.gov

**U.S. Secret Service (USSS)**
245 Murray Lane SW, Building T-5
Washington, DC 20223
202.406.5708
www.secretservice.gov

**U.S. Securities and Exchange Commission (SEC)**
100 F Street NE
Washington, DC 20549
800.732.0330
www.sec.gov

**World Association of Detectives (WAD)**
P.O. Box 20030
Evans, CO 80620
443.982.4586
Wad@Wad.net
www.wad.net

**World at Work**
14040 N. Northsight Boulevard
Scottsdale, AZ 85260
877.951.9191
customerexperience@worldatwork.org
www.worldatwork.org

# INDEX

## C

## D



## E



## F



## G



## H



## I

## L



## M



## N

## T



## U



## W

# The Careers In... Series

Careers in Artificial Intelligence
Careers in Biology
Careers in Building Construction
Careers in Business
Careers in Chemistry
Careers in Communications & Media
Careers in Education & Training
Careers in Engineering
Careers in Environment & Conservation
Careers in Financial Services
Careers in Forensic Science
Careers in Gaming
Careers in Green Energy
Careers in Healthcare
Careers in Hospitality & Tourism
Careers in Human Services
Careers in Information Technology
Careers in Law, Criminal Justice & Emergency Services
Careers in Manufacturing & Production
Careers in Nursing
Careers in Physics
Careers in Protective Services
Careers in Psychology & Behavioral Health
Careers in Public Administration
Careers in Sales, Insurance & Real Estate
Careers in Science & Engineering
Careers in Social Media
Careers in Sports & Fitness
Careers in Sports Medicine & Training
Careers in Technical Services & Equipment Repair
Careers in the Arts: Fine, Performing & Visual
Careers in the Automobile Industry
Careers in the Music Industry
Careers in Transportation
Careers in Writing & Editing
Careers Outdoors
Careers Overseas
Careers Working with Animals
Careers Working with Infants & Children
Careers: Paths to Entrepreneurship